SPECULATIVE TRUTH

RUSSELL McCORMMACH

Speculative Truth

Henry Cavendish, Natural Philosophy, and the Rise of Modern Theoretical Science

OXFORD
UNIVERSITY PRESS

2004

OXFORD
UNIVERSITY PRESS

Oxford New York
Auckland Bangkok Buenos Aires Cape Town Chennai
Dar es Salaam Delhi Hong Kong Istanbul Karachi Kolkata
Kuala Lumpur Madrid Melbourne Mexico City Mumbai Nairobi
São Paulo Shanghai Taipei Tokyo Toronto

Published by Oxford University Press, Inc.
198 Madison Avenue, New York, New York 10016

www.oup.com

Library of Congress Cataloging-in-Publication Data
McCormmach, Russell.
Speculative truth: Henry Cavendish, natural philosophy, and the rise of modern theoretical science / by Russell McCormmach.
p. cm.
Includes bibliographical references and index.
ISBN 0-19-516004-5
1. Heat—History—18th century. 2. Physics—Great Britain—History—18th century. 3. Cavendish, Henry, 1731–1810. 4. Chemists—Great Britain—Biography. I. Title.
QC252 .M23 2003
530'.0941'09033—dc21 2002156309

1 3 5 7 9 8 6 4 2

Printed in the United States of America
on acid-free paper

For Arthur Donovan and Peter Harman

One who proceeds thus far, is an experimentalist; but he alone, who,
by examining the nature and absorbing the relations of facts,
arrives at general truths, is a philosopher.

—William Enfield, *Institutes of Natural Philosophy, Theoretical and Experimental*, 1785

For the creation of a theory the mere collection of recorded
phenomena never suffices—there must always be added a free
invention of the human mind that attacks the heart of the matter. And:
the physicist must not be content with the purely phenomenological
considerations that pertain to the phenomena. Instead, he should press
on to the speculative method, which looks for the underlying pattern.

—Albert Einstein, lecture at the Berlin Planitarium, 1931

PREFACE AND ACKNOWLEDGMENTS

When I first became interested in the eighteenth-century natural philosopher Henry Cavendish, it was thought that all of his surviving scientific manuscripts were in the possession of his kinsman, the Duke of Devonshire. From that collection came the magnificent edition of Cavendish's electrical papers, prepared and annotated by James Clerk Maxwell,[1] as did its sequel, a selection of Cavendish's other scientific papers, edited by Edward Thorpe and others.[2] Since then, two previously unknown groups of manuscripts have come to light, one from within the original collection; the other, from a different branch of the family.[3] The first is a large group of letters exchanged between Cavendish and his colleagues, which together with his other letters were published as a complete edition of his scientific correspondence in 1999.[4] The other group is a small miscellany of his scientific manuscripts, two of which are valuable—one a notebook recording his late chemical experiments, the other a paper on the theory of heat. To my knowledge, the latter is the only developed mechanical theory of heat from the eighteenth century. Sixty years later, a similar theory became a foundation of classical physics, the molecular theory of heat and thermodynamics. A contemporary physicist, who at my request read Cavendish's manuscript on the theory of heat, wrote to me of his reaction: "I am extremely surprised and impressed by what I read. . . . It seems to me that he got the nature of heat essentially right."[5]

In part, this book is an edition of Cavendish's manuscript on the mechanical theory of heat. As such, it is to be understood as a supplement to Cavendish's scientific papers, brought out as a two-volume set by Cambridge University Press in 1921. The publication of the manuscript serves to correct a longstanding, basic misinterpretation of this natural philosopher. Not only was he an exacting experimenter, he was a subtle theorist as well.

This book is at the same time a study of physical theory in natural philosophy during Cavendish's time, the second half of the eighteenth century. Cavendish's theory of heat and the great branch of learning known as natural philosophy illuminate one another. By investigating what natural philosophy was as a field of research, what held it together, what made it work, and what, from our perspective, it could and could not achieve, this study contributes to our understanding of this still rather obscure period in the history of science. As a reader observed, like sunken Spanish galleons and other nautical wrecks, which nowadays are used to reexamine historical epochs, Cavendish's manuscript provides a fitting point of departure for a journey into natural philosophy. Like journeys of old, this one sets out using old charts in the hope of making a correction or two, perhaps of locating a hidden treasure, and in any case, of seeing the world.

FOR PERMISSION TO use material from Henry Cavendish's scientific manuscripts, I thank the Chatsworth Settlement Trustees.

For their help and encouragement, I thank Dietrich Belitz, Sally Bodnar, Geoffrey Cantor, Robert Deltete, Arthur Donovan, Peter Harman, Ingrid Hofmaster, Bruce Hunt, Alexander Morrow, Joseph F. Mulligan, and Lewis Pyenson.

CONTENTS

SPECULATIVE TRUTH

INTRODUCTION

Speculative Truth

Let me begin with a comment on the title of this book. In the eighteenth century, as today, the word "speculative" could be used to belittle, but normally it was not. In 1797 the *Encyclopaedia Britannica* gave the object of natural philosophy as "speculative truth." The natural philosopher was, at times, the "speculative philosopher" and his work "speculative philosophy."[1]

Likely to be concerned with forces and the properties of matter,[2] a speculative philosopher was one who contemplated nature within a scientific framework, and who undoubtedly also made experiments and observations. His reasoning was called "conjectural," "hypothetical," or "theoretical." A modern scholar observes that Benjamin Franklin successfully combined "speculation and experiment" and that this combination has "characterized the growth of physical science during the last two centuries."[3] This statement is surely correct. Although in the eighteenth century, only around three percent of the papers appearing in the Royal Society's *Philosophical Transactions* were purely speculative,[4] a much larger part necessarily contained speculative aspects.

"Truth," the other word of the title, is encountered frequently in scientific writings from the time. "What is truth"? the empiricist philosopher John Locke asked. His partial answer was that "*real truth*" consists in the agreement of ideas in our mind, and our knowledge that the ideas "are capable of having an existence in nature."[5] The empiricist philosopher David Hume similarly answered that truth is either the "proportions of ideas," as in mathematics, or the "conformity of our ideas of objects to their real existence."[6] Natural philosophers ordinarily were not also philosophers of knowledge, but their writings suggest that they commonly held notions of truth similar to Locke's and Hume's. Other philosophical opinions on the subject were possibly shared by natural philosophers; for example, truth resides in objects.[7] In practice, what natural philosophers usually meant by truth was agreement with the facts, with testimony of the senses, with experiment and observation.

In their pursuit of truth, natural philosophers paid equally close attention to error. The natural philosopher James Hutton explained why. Without error, he said, we cannot know truth. The senses do not err, but we can err in judging information conveyed by them, and be deceived. This is the simplest kind of error. We can also err in judging analogies, similarities, identities, and other conceptual relations, the sorts of comparisons that enter physical theorizing. Indeed, all of our scientific judgments are subject to error, for we are not omniscient creatures. Nevertheless, human imperfections does not affect our belief that we can know the world truthfully. If our principles are supported by ample evidence, and if we reason from them correctly, we can form conclusions without the least doubt of their truth. Should our conclusions about the world turn out to be erroneous after all, with the aid of reason and science we can correct them. Truth and error are absolutes, the two extremes of our judgment. In between, where science is to be found much of the time, there are degrees of probability, with the ever-present object of attaining truth.[8] Natural philosophers spoke of "approximations to the truth," not of "approximate truth."

The author of a book on natural philosophy wrote that truth lies "*hidden* in darkness." "Obstructed by passions, prejudices, habits and vices, causes of *error*,"[9] the search demands all the power that human reason is capable of and inspires the utmost determination. It is the calling of the dedicated natural philosopher. A colleague said of Cavendish that "the love of truth was sufficient to fill his mind."[10]

Question

Near the end of his life, Newton told a nephew, "I do not know what I may appear to the world; but to myself I seem to have been only like a boy, playing on the sea shore, and diverting myself, in now and then finding a smoother pebble or a prettier shell than ordinary, whilst the great ocean of truth lay all undiscovered before me." His biographer, who otherwise emphasizes the domineering side of Newton's nature, regards this profession of humility as another important truth about him, his "naive wonderment."[11] As children, we all ask questions about the world. Newton never stopped. He valued what he knew, of course; he held onto his smooth pebbles with a tight grasp, but he never long diverted his gaze from the ocean, the undiscovered, timeless truths of Creation. He filled his writings with primal questions. What is God? What is light? What is gravity?

Newton said that if he had seen farther than others, it was because he had stood on the shoulders of giants. If we suppose that, in invoking this ancient aphorism,[12] his meaning was not a derogation of his rival Robert Hooke, or not merely that, but also direct, then the giants he had in mind certainly included Galileo and Kepler. In his *Dialogue Concerning the Two Chief World Systems*, Galileo made the scientific question a continuous argument for the new physics and astronomy. On the first day, an interlocutor questions the natural motion of the earth. On the second day, another asks what kind of natural events motion and rest are. Elsewhere he pleads, "I do not quite understand the question." In Galileo's other scientific dialog, *Dialogues Concerning the Two New Sciences*, one of the same interlocutors describes himself as "being curious by nature."[13] Kepler, in his *Epitome of the Copernican Astronomy*, reveals himself as having such a nature. He questions why the sun should be seen as the cause of planetary motion. "How do you prove this?" "What do you oppose to this?" "Could you make things clearer by some example?"[14] Kepler's entire book is developed through questions like these. If always tacitly active in research, the scientific question was given a specific formulation by Newton, a "query," and for a time in the eighteenth century the query was a force in natural philosophy. Questions, queries, and quests for truth were subsumed under the natural philosophers' favorite word for what they were about, "inquiry." *Oxford Universal Dictionary* gives its meaning as an "act of asking or questioning."

Science can address questions as sweeping as any in philosophy. However, it characteristically breaks down big questions into small ones, which can be posed in precise, answerable terms. Galileo's interlocutor can see no benefit in drifting upon the waters of philosophy: "We had better get back to shore, lest we enter into a boundless ocean and not get out of it all day."[15] "Problems" was one of Newton's terms for readily answerable questions, as it was Cavendish's. Newton's "queries" merged grand questions with problems, engaging the range of human curiosity and the energy it releases.

In his foreword to an edition of Newton's *Opticks*, Einstein wrote, "Fortunate Newton, happy childhood of science!"[16] Science as we know it was indeed in its youth in Newton's time. No one knew what this prodigy would grow into, perhaps a responsible adult, perhaps a seer, or a savior, or a monster, or any other familiar or frightening human production. Newton did his best to explain what science was and to chart its course. He wrote out the methods of research, the rules of thinking, and the questions that had motivated him. As an ongoing activity in need of recruits, science profited from Newton's writings about science as well as from his examples of science at the highest level of achievement. Taken together, his accomplishments, pronouncements, and questions conveyed an idea of what science was about. Eighteenth-century British natural philosophers tended to look back on him as the Wise One.

With hindsight, we may question Newton's questions. They were, after all, taken seriously because of the questioner's solid scientific accomplishments, not an altogether convincing reason. To many, they were not even questions, but rhetorical truths or dogmas. From our vantage, the questions can seem wrongheaded, having sent a generation of investigators down the garden path of scientific dead ends. Perhaps we should agree that Newton's scientific importance began and ended with his researches in optics and mechanics, with his exacting observations and his glorious mathematics. His questions, then, were scraps on the cutting-room floor, leftovers after the fabric of the universe had been measured and fitted. There is substance behind these dismissals, but in the eighteenth century Newton's questions were rarely viewed that way. It is a matter of record that they stimulated research, and if it was of uneven quality, it included some that was excellent. We proceed on the assumption that Newton's successors had good reason for appreciating the questioning part of his legacy.

People have a healthy fear of questions. If, over the course of a lifetime, they do not answer a reasonable portion of them right, they fail. Teachers browbeat pupils with questions. Physicians frighten patients. Lawyers bully witnesses. "Question" once meant judicial torture. Fortunately for science, questions also serve as a stimulus to productive work.

With a common goal of truth, questions enter science in two ways. One way concerns accepted knowledge and expresses the critical nature of science: "[W]e ought to call into Question all such things as have an Appearance of Falsehood, that by a new Examen we may be led to the Truth."[17] The other way refers to areas of ignorance or partial understanding and expresses the open nature of science. An experimental paper in the middle of the eighteenth century began, "*Problem*, or question proposed" and ended, "*Solution*, or answer to the question."[18] Questions and answers belong to the raw materials of debate—of claim, counterclaim, and argument—by which science moves ahead.

A good popular history of early twentieth-century physics originally bore the evocative title, *The Questioners*. When the book was reprinted, it was given another title, the shame.[19] The original would make an equally fitting title for our book on natural philosophy. The "holy curiosity of inquiry," the theoretical physicist Einstein wrote,[20] drives science, and its proper form is thoughtful questions. Physics, according to a favorite saying of another theoretical physicist, Niels Bohr, is a way of "asking questions of Nature."[21] Questions of science are, of course, connected to other kinds of questions, such as "career questions."[22] These are examined elsewhere, and are not taken up in this book.

Theories, our central interest here, and questions are intertwined. An electrical researcher wrote that the solution to the "difficult questions" in his field "will depend upon the establishment of a more perfect theory."[23] Cavendish's life as a researcher can be arranged under such questions as, What is heat? To this question, his complete answer took the form of a theory, "Heat."

Theory

Scientific theory has long played an important role in humanity's quest to understand nature. Einstein wrote in the foreword to a translation of *Dialogue* that contrary to the common belief that Galileo replaced the earlier speculative, deductive method of science with the empirical, experimental method, what he actually did was to apply his own speculative, deductive thinking, and to oppose the earlier only when its premises were faulty for his goal was not facts but "comprehension." There can be "no empirical method without speculative concepts and systems," Einstein said.[24] What he recognized in Galileo, he recognized in Newton, a thinker like himself, the "first to succeed in finding a clearly formulated basis from which he could deduce a wide field of phenomena by means of mathematical thinking, logically, quantitatively and in harmony with experience."[25] As Einstein meant it to be, that is a good description of the activity of a theoretical physicist today, and, as we will see, of the activity of Cavendish as well.

Einstein summarized the work of experimental and theoretical research as a form of reciprocal questioning: "Inductive physics asks questions of deductive, and vice versa."[26] Natural philosophers thought of their work similarly. The natural philosopher Thomas Young spoke warmly of British science, which since the seventeenth century had displayed "a certain combination of theoretical reasoning with experience."[27] The natural philosopher John Playfair wrote that "in physical inquiries, the work of theory and observation must go hand in hand, and ought to be carried out at the same time." To put off theorizing until all the facts are in, he continued, is an "excess of prudence fatal to all philosophical inquiry." If prudence had been the rule, the imperfections of science would never have been exposed. In the absence of theory, the amassing of observations was useless.[28] The philosopher Dugald Stewart put the case forcefully:

> Nothing, indeed, can be more absurd than to contrast, as is commonly done, experience with theory, as if they stood in opposition to each other. Without theory (or, in other words, without general principles inferred

> from a sagacious comparison of a variety of phenomena) experience is a blind and useless guide; while, on the other hand, a legitimate theory (and the same observation may be extended to hypothetical theories, supported by numerous analogies) necessarily presupposes a knowledge of connected and well ascertained facts.[29]

By the eighteenth century, physical theory had an impressive reach. The same theory that explained the solar system enabled Cavendish to weigh the world by the mutual attraction of lead spheres. Yet, as his successors were to demonstrate, the "depth and power of theoretical methods" in physics had only begun to be hinted at.[30] In time, the extent and intricacy of theoretical physics came to require the attention of a full-time specialist. In due course, the theoretical physicist's predictions would regularly exceed the capability of existing experimental technology. That is where we stand today.

Indispensable as theoretical knowledge is, exactly what it consists in is far from obvious. Philosophers of science grapple with questions about hypotheses, models, and other parts of theories, about the relationship between theoretical statements and the real world, about theory and truth. Entire schools of thought turn on answers to such questions as, What is the function of theory in science? What is the meaning of "theory"?[31]

"Theory" has a number of common meanings. Conjecture is one of them; another is an assertion about something that cannot be directly perceived; another is a belief constituting a world view. Other meanings are a set of assumptions that explain or predict, a hypothesis that is confirmed by experiment or observation, a natural law, and a statement about the causes of phenomena.[32] Because all of these meanings entered natural philosophy, to begin this study with a definition of "theory" would be to ignore its history. We take "theory" to mean what its users meant by it; specific meanings emerge with examples. For reasons of the same sort, we do not begin with a definition of "natural philosophy."

Natural Philosopher

Discussions in this and the following sections may seem to split hairs, but unless we become familiar with the language of the eighteenth century, we cannot expect to understand its science. The term "natural philosophy" was commonly used in Britain, as was the corresponding term for the person who cultivated it, "natural philosopher," while abroad their counterparts commonly were "physics" and "physicist." By "philosophy," or the pursuit of knowledge, British natural philosophers usually meant their own subject, natural philosophy. Consistent with that shorthand, they referred to themselves as "philosophers," and to their scientific societies as "philosophical" societies. In their texts on natural philosophy, they rarely mentioned a philosopher other than one of their own, and if they did, it was likely to be the father of the inductive method in science, Francis Bacon, who in any case was a kind of scientist. Evidently finding in natural philosophy all they needed for their work, they presented their subject within its own rules of reasoning, methods, laws, phenomena, and authorities. Historians note, however, that British empiricist philosophy originated in natural philosophy, and they suggest that the interaction of natural philosophy and philosophy was ongoing.

For purposes of discussion, we take natural philosophers to be, among other things, persons who had a desire to promote and advance the understanding of nature, made a serious study of natural philosophy, devoted time to its activities, and kept informed on aspects of recent research. We are talking about a decidedly small number of persons. The audience for natural philosophy was limited to begin with, and those who contributed to it greatly more so. Few people had the means, time, inclination, or ability to take active part, and those who did customarily led all kinds of busy lives apart from it. Representing different stations within society, variously educated and loosely bound by interest, natural philosophers formed small circles, which met in homes, coffeehouses, and public rooms of formal societies. Readers should keep in mind that in our characterization of the diverse tribe of natural philosophers, exceptions are the rule. Likewise, unless otherwise indicated, anything I attributed to Cavendish refers to him alone.

One reason we take an interest in natural philosophers is that natural philosophy in the eighteenth century led to science in the nineteenth century, which led to science today, and with it to our world of nuclear weapons, electronic computers, satellite communications, and genetic mapping. We grant natural philosophers an honorable place in this story of remarkable success. We study them with the perfect vision of hindsight. We know how it all came out one, two, and three hundred years later.

Lacking our hindsight, why did people in the eighteenth century become natural philosophers in the first place? We ask of them what Hume asked of philosophers in general, why they have "consum'd their time, have destroy'd their health, and neglected their fortune, in the search of such truths" as might benefit the world.[33] What natural philosophers got out of their search for truth was outwardly paltry, was it not? No form of address or rank came with it. If they so chose, they could earn money, but not a fortune. Professors, lecturers, authors, and consultants received payment, but normally to pursue natural philosophy they made a sacrifice. Their own research was self-supported. By and large, anyone who took up natural philosophy did so out of disinterested desire, the force and nature of which depended on the individual. Although probably no two natural philosophers would have answered Hume's question the same way, they would certainly have included among their reasons curiosity, truth, beauty, God, utility, delight, perhaps fame, and certainly the intrinsic dignity of the study.[34]

Among eighteenth-century British natural philosophers, Cavendish stands out for the sustained intensity of his inquiry into the workings of nature. Simply put, his life was about natural philosophy. First and foremost, natural philosophy was his work. Not ordinarily thought of as an occupation, natural philosophy offered him an activity of a kind that was compatible with his aristocratic position in the wider society. It opened for him a career of public service fully as absorbing as traditional careers in politics and government, the military, religion, law, and medicine. With his career came fellowship. Inordinately shy in public, Cavendish came together with a limited society with which he could make human contact. Inquiring, skeptical, and supremely intelligent, he established all of his meaningful connections with the world through natural philosophy. Drawing its strength from the Scientific Revolution of the previous century, in Cavendish's time natural philosophy was sufficiently developed to make possible a life such as his, one characterized by a single-minded dedication to comprehending the universe. Work, understanding, satisfaction, and association—these things made a complete life for Cavendish. Cavendish enters the history of science as a discoverer of new things about the world, which he was, but his most revealing discovery was inward, although he probably did not think of it as a discovery. It was that science was a world, his world, an originality fated to become a hackneyed idea, the narrow specialist whose veins run with ice water.

The expression "natural philosopher" fell into disuse in Britain in the next century, replaced by our "physicist." More than a choice of words, an era had passed. In our day, persons calling themselves "physicists" receive specialized training. Their studies define "physics" for them, and the degree they earn certifies that they have mastered its knowledge and skills at the entry level. By contrast, natural philosophy was open to any and all who had a true concern, and their careers were correspondingly open to individual interpretation.

This Study

Cavendish's manuscripts tell of a life filled with scientific activity. Much of his private research was important, prompting later scientists to wonder why he did not publish more of it. Yet it would seem that he published nearly everything he intended to. He withheld part of his first published paper, on factitious air, but otherwise he fairly informed the world of his work in pneumatic chemistry. He did not publish many of his experiments on electricity, but he did publish his electrical theory on which his experiments were based, which fully exposed his thinking in electricity. In one major subject only, heat, did he fail to reveal his guiding ideas. We should not, therefore, be surprised that he wrote a theory of heat, or that he planned from the start to publish it. "Heat" is the only complete paper among his manuscripts that he indisputably wrote with the intention to publish.

Upon Cavendish's death, his heir invited colleagues from the Royal Society to examine his scientific papers to determine if there were any "he had prepared & thought fit for printing." After poring over the papers for two weeks, Sir Charles Blagden reported that the search had proven "fruitless."[35] The likelihood is that "Heat" was separated from Cavendish's other scientific papers at the time of the search. For if it had been with the rest, Blagden would have recognized it, and with minimal editing it might well have been published in the *Philosophical Transactions* in 1810, in time to make a difference to physics.

"Heat" enlarges our understanding of Cavendish, and as an illustration of theoretical work in the eighteenth century, it is a window onto natural philosophy. As evidently the only attempt to develop a comprehensive theory of heat on mechanical principles in the eighteenth century, it raises a host of questions. That Cavendish was able to carry his mechanical theory of heat as far as he did is revealing of the possibilities of natural philosophy; that he got no further with it is equally revealing of the limitations of natural philosophy.

Part 1 of this book introduces readers to natural philosophy. Its purpose is to characterize the mental world within which the natural philosopher made his choices of materials for representing nature. The mental world together with the choices are our starting point for understanding how a physical theory was made in the eighteenth century.

If part 1 is indispensable for an informed reading of Cavendish's physical theory, its length, level of discussion, and scope all reflect compromises, and as such it will inevitably incur dissatisfactions. As an introduction to the concepts and theories of natural philosophy, it will interest different sorts of readers differently. Those who are familiar with the subject may think the discussion longer than it needs to be to prepare for what follows, yet not long enough to be a self-contained discussion of natural philosophy. Those for whom the subject is new will certainly profit from part 1, which, if anything, they might find too schematic and abbreviated.

Part 2 provides the historical setting for, describes the nature of, and analyzes aspects of Cavendish's theory of heat. Although much of this discussion is given in a 1988 article and again in a biography,[36] it is repeated here for convenience. The appendix is a first edition of Cavendish's complete manuscript, "Heat." Taken together, the two parts of the book and the appendix help us to appreciate the best efforts of a gifted natural philosopher to explain the actions of heat in the world and, at the same time, to bring clarity to our view of a complex period in the history of science. An inquiry into how physical theory was made in the time of natural philosophy broadly illuminates the larger question, to which in one way or another all of the copious, earlier historical studies of natural philosophy are addressed: *What* was natural philosophy?

British natural philosophy was sufficiently distinct from physics abroad for it to be discussed historically in its own terms. The standard book-length treatment of the subject remains Robert E. Schofield's *Mechanism and Materialism*, a fine study of dominant themes of research in that period. There are several helpful books on individual branches of British natural philosophy, such as chemistry, optics, and magnetism, and historians have industriously traced connections between natural philosophy and the wider culture and society. The indebtedness of the present study to the above work is acknowledged in the notes and again in the bibliography.

This study corrects an error in the history of British natural philosophy. It has been maintained that, "unfortunately, though Britain produced a number of scientists in the eighteenth century competent to carry on the empirical tradition worthily, there was no one to extend the Newtonian method of mathematical physics";[37] that "the general concepts of [Newton's] *Principia* were to be kept in mind and applied whenever possible, but there was no attempt on the part of eighteenth-century experimentalists to build a geometrical *Principia*-like structure."[38] Whatever their merits, these statements overlook Cavendish.

Several objections to this study may be anticipated. The most obvious has to do with the form of the book. It is at once a study of the place of theory in natural philosophy and an edition of a theory of heat. Because normally a book on a topic like this is either a history or an edition, I ask readers to bear with me. I believe that in the end, readers will agree with me that two halves make a whole. A related objection is that the way the material is presented is a generation out of date, against which there is probably no defense. The closest parallel is the older history of medieval science, with its emphasis on documents, a luminous tradition now somewhat in eclipse. Another objection is that although Cavendish reveals great physical insight, his theory leaves loose ends and may not always show him at his best. Whatever his difficulties, they reflect a general difficulty of the time in understanding heat mechanically. His evident dissatisfaction with his work is part of its historical value.

Not exactly ignored, but understated, is the experimental side of natural philosophy, an imbalance implicit in our subject. Another objection is that this study refers only in passing to several important features of natural philosophy. The relations between natural philosophy and philosophical, religious, and political beliefs merit a comprehensive study in their own right. For practical reasons, which can be justified,[39] we will consider natural philosophy insofar as it was a technical field, and whatever else it was, it was certainly that, too. I have tried to include everything that a natural philosopher might call upon in making a physical theory, and everything that readers of this study need to know to understand Cavendish's accomplishment.

A final objection is that this study takes the high ground, overlooking important disagreements between natural philosophers. Indeed, it does not follow changes in natural philosophy over time, of which there were many. For example, an author early on in our survey refers to his "illumined Age" and its ideal of a "*truly natural and rational Philosophy*," and an author at the end of our survey characterizes nature as the seat of "opposing or antagonistic principles in a state of perpetual warfare." Bathed in the light of reason, the first author places us squarely within the Enlightenment; attuned to unceasing strife in the world, the second author refers us to a new sensibility, Romanticism. The rationale for not discussing changes like this is practical. The limitations and, I trust, the strengths of this study follow from its restricted focus, natural philosophy as revealed in the making of a theory of heat.

I close with a caveat: Where I comment on Cavendish's theory of heat with reference to the later development of the energy principle, thermodynamics, and the molecular theory of heat, I do not wish him to appear clairvoyant. I use the anachronistic expression "mechanical equivalent of heat" because I do not see a way around it that is not also clumsy.

Natural Philosophy

Natural Philosophers

Characteristics of a Natural Philosopher

A late-eighteenth-century scientific dictionary explained what distinguished a natural philosopher:

> If we consider the difference there is between natural philosophers, and other men, with regard to their knowledge of phenomena, we shall find it consists not in an exacter knowledge of the efficient cause that produces them, for that can be no other than the will of the Deity; but only in a greater and more enlarged comprehension, by which analogies, harmonies, and agreements are described in the works of nature, and the particular effects explained; that is, reduced to general rules, which rules grounded on the analogy and uniformness observed in the production of natural effects, are more agreeable, and sought after by the mind; for that they extend our prospect beyond what is present, and near to us, and enable us to make very probable conjectures, touching things that may have happened at very great distances of time and place, as well as to predict things to come; which sort of endeavour towards omniscience is much affected by the mind.[1]

By this broad characterization, a natural philosopher had a breadth of comprehension, perceived analogies and other regularities, derived rules that explain phenomena, and predicted the future. Another characterization distinguished the natural philosopher from the multitude by the "accuracy" of his observation, the "precision" of his judgment, the impulse of his "speculative curiosity," and the way he acquired facts with the aid of experiments, which enabled him "to place nature in situations in which she never presents herself spontaneously to view, and to extort from her secrets over which she draws a veil to the eyes of others." Owing to his practice of making experiments and attending to detail, the natural philosopher had a stock of information denied to those whose time was taken up by the necessary vocations of life, and whose perception of facts was normally vague.[2] Yet another characterization distinguished the "philosopher" from the "experimentalist": The latter discovered facts and did nothing else, whereas the natural philosopher examined the relations between facts to arrive at general truths, an activity which required "patient attention, deep reflection, and accurate penetration."[3] As these lengthy characterizations suggest, the concept of a natural philosopher was not obvious.

In their study of nature, natural philosophers were expected to exhibit certain traditional virtues.[4] The inevitable model of virtuous inquirer was Newton, who allegedly did not claim wisdom beyond what his experiments permitted him to assert about the workings of nature.[5] He had demonstrated restraint with particular clarity by not pronouncing positively on the underlying physical cause of the force of gravity. Although there were critics who thought that Newton was simply skirting the hard questions of physics by his disclaimers, by and large natural philosophers openly admired him for his presumed modesty as they did for his evident sagacity. Those who followed his example might settle for a permanent state of expectation instead of total comprehension. Their condition was not despairing; on the contrary, when properly appreciated, imperfection was tonic. They might rejoice that the book of nature was never closed, that much of the world lay before them undiscovered. For these natural philosophers, forbearance was both proper and rewarding.

On the subject of human understanding, natural philosophers were in step with British empiricist philosophers. In his great work *Essay Concerning Human Understanding*, Locke called understanding the "most elevated faculty of the soul." "Searches after truth" were the understanding's peculiar form of hawking and hunting, he said, and whoever was not lazily content with the leavings of begged opinions could not fail to achieve the hunter's satisfaction. By its ability to grasp the meaning of words and ideas and the connections between ideas, human understanding opened the doors to "discovery," to "progress towards knowledge."[6] The natural philosopher was Locke's hunter in one of his guises, the search after truth his appointed mission, and the understanding of nature his ultimate quarry.

Natural philosophers would have agreed with a modern philosopher that the faculty of understanding is recognized for its "ability to fashion scientific explanations," and that "perhaps the most important fruit of modern science is the understanding it provides of the world in which we live, and of the phenomena that transpire within it."[7] Others today might object that science is not about understanding, but about other things, about power and prediction, say. In the eighteenth century, however, power was an article of faith, whereas an understanding of parts of the natural world was a demonstrated fact. It is true that natural philosophers valued prediction, and from Newton on, science contained a positivist strain, but the positivists' insistence that the goal of science is prediction was still down the road. Understanding and truth were prized as well.

Publication

Like most voluntary activities, science was organized to meet a variety of needs: fellowship, encouragement, recognition, exchange of information, and advancement of knowledge.[8] These needs were all addressed by the Royal Society of London. Founded in the late seventeenth century, the Society was the most important scientific organization in Britain, although by the late eighteenth century it was joined by royal societies founded in Edinburgh and Dublin, and by scientific and literary societies founded in various provincial centers in Britain. There were a few specialized societies as well, but for the most part their advent together with specialized journals had to wait for the next century. The three royal societies published general journals of science, a main source of the researches discussed in this book.

In Cavendish's day, original science continued to be published in books, and a researcher of Cavendish's stature would normally be expected to publish one or more books, which might be treatises or collections of previously published papers. Cavendish, however, published only papers in a journal, and a paper was the form in which he planned to publish his theory of heat. In his publishing practice, Cavendish foreshadowed physicists who came after him.

Like other natural philosophers, Cavendish was self-educated in the ways of reporting research. The main source of examples was the century-old *Philosophical Transactions*, issues of which came regularly into his father's house during the time Henry was a student at Cambridge. Beginning in the year he came home from the University for good, his father, Lord Charles Cavendish, served on the committee of papers of the Royal Society. Concerned with the good reputation of the Society, which had been brought into question recently in connection with papers, and with the related issue of scientific standards, the committee passed judgment on every paper appearing in the journal.[9] Like his father, Cavendish would serve on the committee, and his scientific papers in the journal would help allay the Society's concern on both counts.

Around 1760, the year Cavendish was elected to the Royal Society, the *Philosophical Transactions* was filled largely with short papers, reports of observations of regularities such as the average rainfall in Plymouth and of oddities such as a two-headed sheep in Devon. From one paper to the next, there was little agreement on questions, approaches, and relevance of the results. To have carefully made and communicated observations was often enough. Experimental papers were then infrequent, and purely theoretical papers even more so. In the 1780s, the decade, we believe, in which Cavendish wrote "Heat," more experimental work was reported, and between papers there was more of a sense of shared problems and standards.[10] In addition, certain categories of papers were becoming rare to extinct, for example, medical cases, antiquities, mechanical arts, and brief communications on any subject. Most papers by this time dealt with subjects in natural philosophy and chemistry, and the remainder largely with subjects in natural history and physiology.[11] Twenty years later, around the time of Cavendish's last important experiment, his weighing of the world in 1798, the number of papers appearing in the *Philosophical Transactions* was only about half as many as when he began his career, and they were twice as long, tending to large surveys of data or exhaustively analyzed experimental procedures. As if to make the point, Cavendish's paper on weighing the world occupied fifty-eight pages of that journal.[12]

The history of the experimental paper as a genre parallels the history of experimental physics. When the *Philosophical Transactions* was founded in the late seventeenth century, the meaning of "experiment" could be as general as "any made or done thing." The goal of experiment then was usually to discover facts or to resolve a debate, and the argument it supported was usually inductive. By the time Cavendish began his scientific work, experiments were often undertaken to prove a hypothesis or to solve a problem. By the end of his career, experiments would be undertaken to prove a general claim or principle. Along the way, experimental papers grew more argumentative, corroborative, and investigative.[13] When the history of the theoretical paper as a genre is worked out, Cavendish's abortive paper on the theory of heat will be a useful, early marker.

The pattern of authorship in the *Philosophical Transactions* reflected the level of complexity of research. Nearly all papers appearing there had a single author, who reported on his individual efforts. An experiment might require a second person to turn the electrical machine, but he was usually an assistant, who was not an author and was often not named in the publication. Instruments were simple, usually made to be operated by a single observer; apparatus was ordinarily simple, too, in keeping with the simple plan of experiments then. When on occasion experiments were carried out by more than one person, curiosity might be the reason, or a need for witnesses. An instance was an experiment on the communication of an electrical shock over a long distance, here clear across the River Thames. Before it was over, the experiment engaged some twenty-five members of the Royal Society, although not that many were needed, as the experiment was also regarded as an outing and an object of amazement.[14] Researchers who collaborated on scientific projects usually did so by appointment of the Royal Society. Weighing the world by the attraction of a mountain was a project of this sort, as was the measurement of the distance between the Greenwich and Paris observatories. Theoretical research seems always to have been carried out by one person working alone. For purposes, of this book keep in mind that Cavendish undoubtedly wrote his paper on the theory of heat for publication in the *Philosophical Transactions.*

Physicists

British natural philosophy had a European setting. When Cavendish worked out his mathematical theory of heat, probably the late 1780s, Paris was the undisputed center of mathematical physics in Europe.[15] We note three pertinent developments in the rise of French physics from the second half of the eighteenth century. One was the institutional recognition of physics, notably by the Paris Academy of Sciences. The second was the mathematical development of physics, which also had an institutional base, especially in the Academy and in the military academies. The third development was a direction of physical research associated with the mathematical scientist P. S. Laplace. Joining mathematical theory and quantitative experiment, Laplacian physics looked for the explanation of all physical phenomena in the action of attractive and repulsive forces between particles. Although in methods and goals British natural philosophy differed from French physics, there was no one distinction between them, but rather differences of degree. British institutions did not designate the boundaries of physics as sharply as did the French, although British universities had professors of natural philosophy who taught the subject separately from natural history and chemistry, and Cambridge University emphasized the mathematical side of natural philosophy. British natural philosophy did not exhibit as much coherence as did French physics during the Laplacian phase, but it also explained phenomena on the basis of attractive and repulsive forces.[16]

To look ahead, Laplacian physics was followed in the early nineteenth century by S. D. Poisson's theory of electricity, A. M. Ampère's theory of electrodynamics, A. J. Fresnel's theory of light, and J. B. J. Fourier's theory of heat, to name only the better known theories to come out of France. If lacking the earlier conceptual unity of the Laplacian approach, this work gave to French physics a uniformly impressive mathematical stamp. Upon its reception in Britain, it transfused "new blood into British physical science."[17]

In Britain, the word "physics" came into common usage around the time that the subject began to acquire attributes of a learned profession, with institutional recognition coming in the 1830s in the form of a special section for physics and mathematics in the British Association for the Advancement of Science and a committee on physics in the Royal Society of London.[18] Physics can be said to have come into being in Britain in the nineteenth century, when disciplinary organization, mathematical and experimental methods, and physical concepts came together to give physics a decidedly different look. To physicists then, earlier men of science, Cavendish definitely, but even those coming after him such as Humphry Davy and Michael Faraday, seemed like figures from a bygone era.[19] Yet there is a valid, if limited, sense in which we can speak of physics in Britain in the nineteenth century as continuous with natural philosophy. Not everything about physics was new.

For a moment, let us adopt a long perspective. Independent of research grants, having to satisfy no employer, and under no necessity to publish, Cavendish looks quaint indeed alongside today's physicists. Working alone in his laboratory, rooms fashioned from the living space of his suburban villa, he presents a strange contrast to today's world, in which parts of physics are given over to teams consisting of theorists, experimentalists, instrument makers, and experts of the kind familiar in the business world—managers, coordinators, and group leaders—who swarm around giant research machines, carrying out projects with public support.[20] Yet, as we will see, the questions posed by natural philosophers and those posed by today's physicists have a common thread.

Philosophies

Philosophy, Qualities

Influential as he was in his lifetime, Locke dominated the intellectual life of the next century to an even greater degree.[1] Certain of his teachings are germane to our topic. According to Locke, the mind perceives the world indirectly, through "ideas," and the powers of bodies to produce ideas in the mind he calls "qualities." Ideas of size, shape, solidity, number, and motion are simple ideas of "primary" qualities; they belong to material bodies, they have an objective existence. By contrast, "secondary" qualities such as taste, smell, color, sound, and heat do not reside in the bodies themselves but are powers that arise from the primary qualities and depend on the mind for their existence.[2] Although the distinction between primary and secondary qualities did not originate with Locke, it was commonly identified with him, and his formulation of it had significance for natural philosophy. Cavendish's mathematical theory of heat was Lockean in that it explained heat, a secondary quality, as a power arising from a primary quality of matter, motion.

Locke's distinction came to be challenged in the eighteenth century. Hume thought that we can be certain of only one reality, our sensations and reflections.[3] George Berkeley thought that no more than secondary qualities do primary qualities exist in external bodies, that nothing exists but the Deity. James Hutton believed that primary qualities are not Locke's simple ideas, but the product of reflection.[4] Whatever their position on this question, Cavendish and other natural philosophers often used the word "quality," usually with reference to motion or to some other property or power of matter.

On the existence of an external reality behind the qualities, natural philosophers rarely commented. In lectures delivered in Trinity College, Dublin, the

Professor of Natural Philosopher said that in natural philosophy, "reality depends on the reality of our sensations; and is not therefore affected by the existence or non-existence of an external, material world."[5] On this issue, most natural philosophers probably sided with Hutton, who believed that objects have a reality.[6]

Natural Philosophy

Natural philosophy began, according to histories of Greek science and philosophy, with the Milesians, who proposed, criticized, and debated theories about the material substance of the world, and who gave natural explanations of phenomena. As we pick up the inquiry into nature two millennia later, as it emerged from the Scientific Revolution, "natural philosophy," and its synonyms "physics" and "physiology,"[7] in their broadest meaning, stood for the study of the entire natural world, material and immaterial. Insofar as it applied to the material world, in scope natural philosophy corresponded more or less to our natural science, although the word "science" had other meanings in the eighteenth century.[8] Ignoring ambiguities of the word "natural,"[9] a scientific dictionary defined "natural philosophy" as "that science which considers the powers of nature, the properties of natural bodies, and their actions upon one another."[10] Most natural philosophers, I think, would have agreed that this definition was close enough.

Of historical interpretations of British natural philosophy, that which has been studied with greatest thoroughness is theories of matter originating with Newton's writings. This interpretation offers a reasonable orientation to most texts on natural philosophy, but it has limitations. It assumes or implies an identifiable, more or less unified Newtonian tradition, the existence of which has been questioned, and only with difficulty, if at all, does it address non- or anti-Newtonian writings on natural philosophy, with their varied motivations, religious, political, and otherwise. This difficulty was acknowledged early on: The author of a text on natural philosophy complained that those who worshipped Newton branded critics of Newton the "worst of heretics."[11]

Other interpretations of natural philosophy characterize it by its audience, by notions of scientific discovery, by public lectures, and by instruments; liken it to drama and entertainment; and identify it with pre-science or an intermediary between philosophy and science. It is called a "community" with a claim to represent nature, a "discipline," a "discourse," a "taxonomy," a "world view," and, least confining of the characterizations, a "practice."[12] If natural philosophy is taken to be what natural philosophers practiced, one description of it is the work that Cavendish carried out during his fifty-year career: He made experiments, observations, and theories in physical science; he designed, built, and used apparatus and instruments; he served on committees for the national scientific society; and he wrote up his researches for himself, his colleagues, and publication. Because of his social position and independent means, in private life Cavendish was unusual among natural philosophers, but in science he was not all that unusual. A number of his fellow natural philosophers did exactly the same kinds of things he did. We will limit our discussion here to the study of nature that was called "natural philosophy."

Commonly, if not always, natural philosophy was distinguished from natural history. The distinction was based on a fundamental division of knowledge: philosophy was knowledge of the nature and reasons of things, history, knowledge of bare facts. Accordingly, natural philosophy studied the powers residing in bodies; natural history described the motions caused by them.[13] According to another distinction, the former studied the properties of bodies; the latter, the classification of specimens. The division was not sharp; butterflies were classified, but so were conductors of heat and electricity. In their fact-gathering activities, experimental fields such as heat and electricity depended upon careful classification.[14] In addition to crossing over, the two categories came together in fields that depended equally on natural philosophy and natural history. Geology, in its fundamentals, necessarily connected with "almost every branch of the physical sciences," in particular, with "precise points of natural philosophy and natural history."[15] Cavendish, who worked on heat, electricity, and other fields belonging to natural philosophy, also worked on geology, mineralogy, and meteorology, fields with a strong component of natural history.

Natural philosophy was commonly distinguished from chemistry, too. One way was to associate natural philosophy with mechanical, mathematical laws, and chemistry with laws of a different order, not reducible to mathematical calculation. One text limited natural philosophy to subjects concerned with visible motions of matter, excluding chemistry, which presupposed invisible motions.[16] In whichever manner natural philosophers and chemists distinguished their subjects, they tended to agree that in some cases, "no boundaries can be established between them."[17] A case in point was heat, the subject most completely incorporated into both. In lectures and texts, heat was often included under chemistry, but it was increasingly included under natural philosophy, too.[18]

In the last edition of the eighteenth century, the *Encyclopaedia Britannica* explained that it was efficient for university professors of natural philosophy to confine themselves to mechanical topics and to a few other subjects such as electricity and heat, since chemistry and physiology had their own professors.[19] Beyond the walls of the university, authors had more discretion. A text on natural philosophy shortly after the turn of the nineteenth century noted that geology, botany, and various other sciences could be arranged under natural philosophy, but because of the quantity of recent discoveries, it was beyond the "capacity of one man" to comprehend them all, an argument from convenience for boundaries in this case. This text covered what would become known as "physics," as well as parts of meteorology and chemistry, but not astronomy because it had its own literature, again a pragmatic reason; the subjects included were, above all, those that interested the author as a researcher.[20] Without general agreement on the scope of natural philosophy, varying interpretations entered research, teaching, and the conceptual organization of science.

Early in the century, topics Newton had treated mathematically—mechanics, hydrostatics and pneumatics, gravitational astronomy, and optics—were a de facto definition of natural philosophy, a narrowing of the subject relative to past understandings. Later in the century, other parts were regularly included, widening its domain again. In the 1730s, a text listed seventeen "Principal Phaenomena of Natural Philosophy," those of electricity, magnetism, and heat not among them.[21] As these subjects gained prominence, texts on natural philosophy increasingly came to resemble later ones on physics. This was the case in Cambridge, where in the second half of the eighteenth century, the Professor of Astronomy and Experimental Philosophy Anthony Shepherd made the main divisions of his course the same as Newton's mathematical subjects, although within the division "Mechanics" he included electricity and magnetism.[22] Electricity and magnetism soon became divisions in their own right, a recognition of their independent importance. Lectures on natural philosophy given by Shepherd's successor, Samuel Vince, and lectures given at the same time by the college lecturer George Atwood contained identical subjects, together with Newton's subjects, they included magnetism and electricity under separate headings.[23]

In addition to the above divisions, natural philosophy had an internal division, a foreshadowing of the division of physics into experimental and theoretical halves, of interest to this study. According to a distinction of the time, natural philosophy consisted of an "experimental" part, which investigated, and a "mechanical" part, which explained.[24] According to another, experimentalists viewed nature as "*disunited*"; systematic writers, as "*united*."[25] By yet another distinction, "men of fact" valued the "practical department" of science, content to register one fact after another, whereas "men of theory" regarded "general results as the great and dignified objects of Science," valuing facts only as they illustrated general theorems. "Enlightened" men of science avoided these extremes by cultivating and combining both theory and fact.[26] By this standard, Cavendish was an enlightened natural philosopher.

In their books on natural philosophy, authors did not belabor the classification of knowledge, but proceeded directly to the elements of their subject. Under headings such as "Introduction to the Study of Natural Philosophy," they stated Newton's method of experimental philosophy and his rules of reasoning, which enjoined natural philosophers to reason inductively and to observe the economy and analogy of nature;[27] to Newton's rules, they might add basic "axioms of philosophy," such as, Nothing can be produced from nothing, or, Nothing is without properties.[28] In the same place, they laid out the basic notions of motion, force, and the properties of matter. Readers could come away with the not unreasonable conclusion that natural philosophy was a way of thinking about the physical universe.

Beauty and God

To the question of why we can know nature, physical theorists today may answer, "Because nature has a simplicity and therefore a great beauty."[29] They may acknowledge that in deciding between theories, aesthetic taste sometimes plays a part alongside experiments, principles, and analogies to other theories. Natural philosophers were not very different. They believed that nature has an underlying simplicity, which might or might not be an aesthetic value. Even an explanation by Newton could be faulted for not being "consonant to that Simplicity, Uniformity, and Regularity, with which Nature is every where observed to act."[30] At the same time, natural philosophers were cautioned not to "conceive a greater simplicity in nature than there really is."[31]

Because "every thing in nature is beautiful," James Hutton wrote, natural philosophy gives rise to a "taste for beauty," the "beauty of order."[32] Natural philosophy reflected the beauty of its subject, its imperfections seen as a kind of beauty mark. Newton was likened to the fourth-century painter Apelles, whose unfinished works were preferred by the ancients over finished works by other painters.[33] Adam Smith compared the natural philosopher's sensitivity to the orderly connection of nature with the musician's sensitivity to the consonance and rhythm, or dissonance and ill-timing, of notes:

> As in those sounds, which to the greater part of men seem perfectly agreeable to measure and harmony, the nicer ear of a musician will discover a want, both of the most exact time, and of the most perfect coincidence: so the more practised thought of a philosopher, who has spent his whole life in the study of the connecting principles of nature, will often feel an interval betwixt two objects, which, to more careless observers, seem very strictly conjoined. By long attention to all the connexions which have ever been presented to his observation, by having often compared them with one another, he has, like the musician, acquired, if one may say so, a nicer ear, and a more delicate feeling with regard to things of this nature. And as, to the one, that music seems dissonance which falls short of the most perfect harmony; so, to the other, those events seem altogether separated and disjoined, which fall short of the strictest and most perfect connexion.[34]

By discerning an imperfect connection, a disharmony, in nature, the natural philosopher began to question, and then he went to work.

Alongside beauty, the Deity entered discussions of natural philosophy. The Professor of Astronomy and Geometry in Cambridge made the familiar connection between the two, telling his students that the starry sky "gives beauty to the creation, displays to us the wisdom and power of the creator."[35] God was the "Great Designer," the "divine Architect," who endowed His creation with symmetry, elegance, and beauty, qualities that were externally obvious to the unphilosophical spectator and internally so to the philosophical.[36] Responding to the beauty of nature, devout natural philosophers could regard their work as an observance. By revealing the laws of God's design for creation, they strengthened the religious convictions of all.

Historians of science look upon natural philosophy as having developed within a "larger debate" on the relation of God and nature, involving theological controversies, arguments over creeds, and church politics. In this debate, natural philosophy had its critics and its champions. The former complained that particles and forces were a form of materialism, and that the experimental method denied revelation, a danger to Church and state.[37] The more common opinion was that natural philosophy supported religion.

On the relation of natural philosophy to religion, Newton had thought long and hard. He discussed God in the General Scholium of the *Principia*, and in *Opticks* he wrote, "If natural Philosophy in all its Parts, by pursuing this [Experimental] Method, shall at length be perfected, the Bounds of Moral Philosophy will also be enlarged."[38] He believed that a perfected natural philosophy would bring people closer to God and to their duty to Him and to one another.

A mid-eighteenth-century account of Newton's work agreed that natural philosophy was subservient to natural religion and moral philosophy. Natural philosophers sought the "true constitution of things," the "real state of things," aware that "false" natural philosophy led to false religion and atheism. The "most noble pursuit" of natural philosophy was to trace the chain of causes, for every cause is the effect of a prior cause until the chain comes to an end with the first cause, God. This "great mysterious Being" is beyond sense experience, its nature and essence "veiled in darkness," "*unfathomable*."[39]

If not always in agreement with Newton, or with one another, eighteenth-century natural philosophers held firm views on God and nature. Citing Newton to justify the discussion of religious matters in natural philosophy, George Adams, instrument maker to the king, an admirer of Hutchinsonian religious writings, published a text intended to show that when properly conceived, natural philosophy was fully compatible with revelation.[40] Joseph Priestley, a Dissenting minister, believed that the powers of bodies derived from God. James Hutton, a deist, believed that nature was self-sustaining, without need of ongoing help from God, and that the laws of nature were immanent in the world.[41] Perhaps in light of his dual occupations, the former Professor of Chemistry in Cambridge, now Anglican Bishop of Llandaff, expressed his thoughts on God and nature at unusual length for a paper in the *Philosophical Transactions*:

> For though the line of human understanding will never fathom the depth of divine wisdom, displayed in the formation of this little globe which we inhabit; yet the impulse of attempting an investigation of the works of God is irresistable; and every physical truth which we discover, every little approach which we make towards a comprehension of the mode of his operation, gives to a mind of any piety the most pure and sublime satisfaction."[42]

With humility, the Christian philosopher upheld the experimental investigation of nature.

Sufficiently confident of their subject, writers on natural philosophy rarely felt the need to justify it. Books on natural philosophy might omit God entirely without risk of impiety. Thomas Rutherforth, who in a few years would become Regius Professor of Divinity in Cambridge University, lectured on natural philosophy without mentioning God. His reason for denying that gravity was an essential property of matter was logical, not religious.[43] Cavendish made no reference to God in any of his publications or manuscripts.[44]

Matter

The primitive concepts of natural philosophy are space, time, matter, and force or other action. The first two, space and time, Colin Maclaurin, Professor of Mathematics in Edinburgh, wrote, are the most clearly conceived quantities.[45] Their parts are uniform: Absolute time flows uniformly; absolute space is the immovable, everywhere similar, infinite emptiness. With respect to absolute space, a body has an absolute location and motion; absolute motion is a reality, as proven by centrifugal force, although on this point there was dissenting opinion. In addition, a body has a relative location with respect to other bodies, and a corresponding relative motion.[46] Texts on natural philosophy usually treated space and time briefly, as concepts needing little elaboration. They did not always make that assumption about the other two fundamental concepts, matter and force.

Matter was assumed to be composed of extremely minute particles. We have no sensation of these particles individually, not because of their nature but because our senses are not "*accurate* enough."[47] Particles in the aggregate, so our reason tells us, cause our sensations, the phenomena of nature.

A text on natural philosophy introduced mechanics with the readers' purported first question, What is matter? Matter constitutes bodies, the "*physical world*," the answer went, and if its underlying substance is unseen, its modifications, qualities, or properties are the object of our senses, "*and hence also of physics*."[48] Of the choices, "properties" was the term most often used in discussions of matter in the eighteenth century.

Newton attributed to matter universal, invariant properties: hardness, extension, impenetrability, mobility, and inertia. These same properties were ascribed to particles, but because they could not be observed directly, they were reasoned about by analogy and an extension of the rules of induction. Not "essential" to matter, the properties were known only by experience; matter without, say, hardness was conceivable. Regarded as primary causes, the properties of matter were discussed and debated through the eighteenth century. Authors freely added to and subtracted from Newton's list, allowing that still other properties may have escaped their notice, or that humans may lack the requisite sense organs to perceive them.[49]

Whether or not gravity is a property of matter was a hard question. As a universal force proportional to an invariant magnitude, quantity of matter, gravitation would seem to qualify as a property of matter; by virtue of being inessential to matter, as Newton regarded it, gravitation would again seem to qualify. The issue was not clear-cut, however, and natural philosophers were divided; some included gravity among the properties of matter, some not, while others included all attractions and repulsions. Scholars today differ in their reading of Newton on the status of gravitation.[50]

Texts on natural philosophy often described matter as passive, as Newton did. An alternative view, which gained adherents as the century progressed, was that matter is active, known by its powers, Locke's qualities, or Newton's attracting and repelling forces.[51] Within the broad interpretation of matter as alternating spheres of attraction and repulsion, there was a difference of opinion about what lies at the center, hard atoms or points. The natural philosopher John Michell held the latter opinion, as did Priestley, although the two differed on another issue. Michell spoke of forces as "properties," which, Priestley said, presupposed a substance in which they adhere, and Priestley could form no idea of a substance which itself is devoid of all properties. Upon this rigorous affirmation of the empiricist philosophy, and supported by the prevailing belief in the near vacuity of the universe, Priestley proposed that "matter consists of *powers* only, without any substance."[52]

By resolving matter into attraction and repulsion, Priestley explained the independent property of solidity, hitherto the essence of matter, as an effect of the power of repulsion, and he explained the property of inertia as well. Against his reasoning, it was argued that inertia and solidity were indispensable, that inertia was the foundation of Newton's laws of motion and thus of natural philosophy, and that unsolid matter was a violation of the universal maxim that a thing cannot act where it is not. Lacking proof, the idea of the penetrability of matter remained a debatable hypothesis. Some authors, Cavendish among them, agreed with Priestley that solidity is an effect of the repulsive force of matter, while others agreed with Newton that it is an original property of matter, and still others held that it is an effect of the aether. In general, opinion on the properties of matter remained unsettled through the end of the eighteenth century.[53] Matter proved a messy subject.

Another basic question was whether there is a single matter of the universe or various kinds of matter. Newton had needed only "One Sort of Matter" to explain the "infinite Variety of Bodies" of our experience, Benjamin Martin wrote in his book on the Newtonian philosophy.[54] However agreeable to the simplicity of nature and the unity of natural philosophy, the idea of a single matter of the universe came into question. Natural philosophers had experimental grounds for thinking that there were probably several kinds of matter, including the four hypothetical, imponderable fluids of light, heat, electricity, and magnetism, and maybe a fifth, phlogiston. They understood that these fluids were distinct, incapable of being transformed into one another, possessing some but not all of the properties of ordinary matter. They also considered another kind of hypothetical fluid, universal in its action, the aether, commonly referred to as "Newton's aether."[55]

Newton regarded the properties of matter as the foundation of natural philosophy. His statements on the subject were closely studied by his successors, who were puzzled by them, as are today's historians and philosophers, who wonder if a consistent natural philosophy can be built upon them. Ideas about the constitution of matter influenced research, and if they did not drive it in the way that experimental apparatus, instruments, and techniques did, they entered essentially into the understanding of what constituted natural philosophy.

Causality, Force

A scientific investigation might result in a bare description of regularities of phenomena, or in an explanation of regularities from a physical cause; the latter was usually considered the deeper of the two and, according to one opinion, the only kind of explanation.[56] A flat statement appearing in several publications of the time was that natural philosophy investigates the causes and effects of bodies.[57] Typically, Thomas Hornsby began his lectures on natural philosophy in Oxford University: "Natural Philosophy is that part of science which examines the different Phenomena of nature—enquires into their causes & traces their effects."[58]

The founders of British empiricism, Locke, Berkeley, and Hume, all made important observations on causality. Locke equated "cause" with "substances" endowed with "the source from whence all action proceeds," or "powers,"[59] a common meaning in eighteenth-century natural philosophy. A weak meaning of "cause" as lawful succession was favored by the Scots, as it was by Berkeley, who otherwise maintained that natural philosophy was about laws, not causes.[60]

The entire notion of causality was subjected to a penetrating analysis by Hume.[61] "All reasonings concerning matters of fact seem to be founded on the relation of *Cause* and *Effect*," Hume said, and "by means of that relation alone we can go beyond the evidence of our memory and senses."[62] Hume did not doubt that physical objects give rise to our perceptions, nor did he seriously question the world view of Newton and Locke, but he did question our justification for believing in these things. He attributed our belief in cause and effect to a habit of thought arising from the constant conjunctions and temporal sequences of our perceptions. Science is based on our feelings, not on our reason.[63]

Closer than Hume to natural philosophers were the Scottish Common Sense philosophers. Thomas Reid regarded Hume's philosophy as a reductio ad absurdum of empiricism. Hume, by casting doubt on the legitimacy of the experimental method of induction, on the assumption of the conformity of nature to itself, and on external reality, seemed to deny Newton's rules of reasoning in natural philosophy. Reid set about to restore common sense to philosophy by laying down inborn first principles, incapable of proof. In this spirit, he posited our belief in causality and in the reality of a material world endowed with primary qualities.[64]

Yet on the subject of causality in nature, Reid expressed himself similarly to Hume. He believed that causes do exist in nature,[65] but that it was not the business of natural philosophy to discover causes regarded as powers to produce effects. When we say "that one thing produces another by a law of nature, this signifies no more, but that one thing, which we call in popular language *the cause*, is constantly and invariably followed by another, which we call *the effect*"; because we have no idea of their connection, all that natural philosophy can do is to trace the laws of nature by induction from phenomena, following Newton's rules.[66] His philosophical colleague Dugald Stewart, who agreed that we perceive no connection between things, identified the work of natural philosophy with the determination by experiment and observation of constant "conjunctions of successive events, which constitute the order of the universe."[67]

After Berkely, Hume, and the Common Sense philosophers, when speaking of causality natural philosophers had ample reason to be circumspect. They had no need to speak of "causes" at all, but only of constant forerunners of events and of the general laws that incorporate them. In their writings on natural philosophy, however, they largely retained the less cumbersome "causes," treating causality as a principle without need of philosophical justification or experimental confirmation.

A familiar cause was a force or power acting on, inhering in, or constituting some kind of matter. To say that one phenomenon "causes" another was an alternative way of saying that the latter depends on "some force or power" in the former.[68] A common formulation in the eighteenth century, supported by Newton's statements of the laws of motion, was that force or power is the cause of the change of state of a body with respect to rest or motion, or of resistance to that change.[69] The first law of motion affirms the persistence of a state of rest or uniform motion in a body in the absence of impressed force; the second law relates change of motion in a body to an impressed force;[70] the third law states that the mutual actions of two bodies are equal and oppositely directed. In the statement of this last law, Newton spoke of "action" and "reaction" instead of "force," and it would seem that he distinguished between the two. Today we read his third law, like his first and second laws, as referring to "force," as did Cavendish.[71] Force was an everyday experience, which Newton illustrated by the pressure of a finger against a resisting stone.[72]

The investigation of forces had three parts, Newton said. The first, which belonged to "mathematics," was the determination of the quantities of forces in general. The second, which belonged to "physics," was the comparison of these quantities with the phenomena of nature to decide which laws obtained in the world. The third was the investigation of the "physical species, causes and proportions" of the forces.[73] The propositions of the *Principia* are concerned with the first two parts. The third is discussed in the General Scholium of that book and in the queries of *Opticks*, where Newton suggested that an active principle, the aether, was a possible cause of gravitation and of other attractions and repulsions. If uncertain of the cause of forces, natural philosophers could still advance their subject, but to that extent natural philosophy could be seen as incomplete.

Newton recommended his investigation of gravitation as a model for the investigation of other forces, by which they, too, might one day become principles. His successors accepted the model, but understandably found it hard to follow.[74] The difficulty, which was inherent in the project, was compounded by a variety of current meanings and associations of "force," reinforced by Newton's different statements on the subject.[75] Ordinarily, writers on natural philosophy did not dwell on ambiguities of "force," but instead directed their attention to laws of force. Newton had shown what could be learned about nature from the one mathematical law of force he had determined, gravitation. Anticipating like rewards from laws of other forces, down the century his successors made hopeful experimental measurements of the forces of electricity, magnetism, and cohesion.

In the year Newton died, Stephen Hales published an account of experiments on air fixed in bodies and capable of being freed, so-called "factitious air."[76]

His explanation of the experiments drew attention to the queries of *Opticks* and, in particular, to the discussion there of forces of repulsion. Other authors had used repulsion before Hales, but he was the first to recognize that both repulsion and attraction were needed to account for the balance of nature. From that time on, natural philosophers invoked repulsive forces with the same confidence they did gravitation, the archetype attractive force.[77] Citing Hales in support, the experimental demonstrator of the Royal Society J. T. Desaguliers stated the need for both attractive and repulsive forces to explain the elasticity of air, strings, springs, and steel balls: "Attraction and Repulsion seem to be settled by the Great Creator as first Principles in Nature; that is, as the first of second Causes; so that we are not solicitous about their Causes, and think it enough to deduce other Things from them." To get on with their work, natural philosophers needed only know that attraction and repulsion exist.[78] With this confidence, in 1748 the natural philosopher Gowin Knight published an explanation of all natural phenomena on the basis of two active principles, attraction and repulsion, and two corresponding kinds of matter.[79]

Natural philosophers recognized several kinds of force. To begin with, there was the sustaining, resisting, and impulsive force, which Newton called *vis insita*, or "inertia (*vis inertiae*) or force of inactivity," which became standard usage.[80] We call it "inertial mass," but Newton seems to have regarded it differently, as a true force, distinct from mass and motion.[81] His successors might follow Newton's designation of it as a force,[82] or they might object to calling inertia a force on the grounds that no force is required to keep a body in motion.[83]

Newton identified three types of impressing forces, percussion, pressure, and centripetal force such as gravity; he also pointed to possible ways to join them. He analyzed centripetal force as percussion, as a succession of tiny blows, or instantaneous contact forces. He also illustrated centripetal force by a stone whirled in a sling, identifying the force with the pull directed toward the hand, and he regarded bodies revolving in orbits such as planets the same way, suggesting that centripetal force and pressure might be regarded as the same force. Differing with Newton on the structure of matter, some British natural philosophers inverted his analysis, regarding pressure as fundamental and collisions as a succession of tiny pressures, a continuously acting contact force; and by analogy with a thread, exerting a continuous pressure on a body, moving it from rest and gradually accelerating it, they regarded the forces of cohesion, gravity, electricity, and magnetism as exerting pressures. There was, then, possibly only one instead of three kinds of impressing force, a step in the direction of the economy of causes and the unity of natural philosophy.[84]

In texts on natural philosophy, the impressing forces normally consisted of five attractions, and more or less the same number of repulsions. The attractions were those of gravity, cohesion, electricity, magnetism, and chemistry, the latter usually distinguished into several varieties, to which might be added the attraction of life, or muscular contraction.[85] Repulsions were paired with attractions: antigravity or levity, anticohesion and elasticity, electricity, magnetism, and chemistry, although others such as the shrinking of leaves of sensitive plants upon touch or the repulsion that keeps stars from falling together might also be included. The anticohesive force, called simply the "repulsive force," keeps particles from colliding or coming infinitely close to one another; the antigravitational force remained controversial. Like Newton, on the basis of the analogy of nature, natural philosophers acknowledged that still other forces might exist. The forces of the elementary particles of matter, as manifested in chemical processes, seemed "almost infinitely various" and might never be known.[86] By any reasonable count, the number of impressing forces was large and probably growing.

There was nothing to prevent a natural philosopher from postulating an impressing force for each distinct group of phenomena; in this way he could get on with his work without having to address the question of first principles. The drawback was that the true simplicity of nature might be lost sight of. Helpful once again, Newton proposed two ways of bringing the forces of nature under a unified viewpoint, both of which had consequences. One was to posit an all-pervasive medium consisting of fine, mutually repelling particles responsible for all impressing forces. This proposal gave rise to an enthusiasm for aetherial explanations from the middle of the eighteenth century; Bryan Robinson made a case for a certain kind of elastic aether as the cause of the attractive and repulsive forces associated with gravity, light, electricity, heat, elasticity, cohesion, fermentation, and muscular motion and sensation.[87] The admission of an elastic aether like this did not eliminate forces, but by reducing all attractions to repulsion, it offered natural philosophy the promise of a great "simplification."[88]

The other way was to conceive of the several forces of nature as though they were a single force. This was suggested by Newton who likened the alternation of attractive and repulsive forces to that of algebraic numbers passing between positive and negative values.[89] Guided by Newton, and also by G. W. Leibniz, with the object of explaining all physical phenomena, R. J. Boscovich developed a theory of natural philosophy based on a single law of attractions and repulsions, represented by a continuous curve looping above and below an axis passing through the force center.[90] Boscovich was not British, but his ideas on forces had a substantial British following, which included Cavendish.

There is "*no insulated fact in nature*," one author observed, and it is the business of natural philosophy to show us why the infinite facts of nature seem separated when they are actually connected, and thereby "lead us to that principle of *unity* which harmonizes, and connects all the works of *creation*."[91] Playfair asked if there might not be a principle more general than the laws regulating the phenomena of gravity, light, heat, electricity, and magnetism, a principle that connects the singular force of universal gravitation to all of the other forces. He answered, or prophesied, that if such a "great principle" exists, its discovery might come in a future age, when "science may again have to record names which are to stand on the same levels with those of Newton and La Place." Concerning "such ultimate attainments," he added, it was "unwise to be sanguine, and unphilosophical to despair."[92]

This way of thinking laid the groundwork for the physics of the interconversion of forces and the conservation of energy, developments dating from the first half of the nineteenth century; beyond that, it set a basic problem for physics for the next two hundred years. Today, there are roughly the same number of forces, or interactions, as there were then, and there are the same expectations. Physicists correctly see themselves as upholding a long tradition, which originated with Newton's wish to see all of the phenomena of nature derived from a few general mechanical principles and forces, and which continued on in Einstein's effort to unify the electromagnetic and the gravitational forces. They still seek Playfair's grand unifying principle or principles, and through them a "unifying theoretical basis" for all of physics.[93]

Experimental Philosophy, Newtonian Philosophy, Mechanical Philosophy

We may find it hard to keep separate the several "philosophies" of science in the eighteenth century, and for the reason that, at the time, they were not always kept separate. The first sentence of Desaguliers's text on "Experimental Philosophy" gave a definition not of that term but of "Natural Philosophy."[94] "Natural philosophy" and "experimental philosophy" were used interchangeably, although they also had their separate characterizations.[95]

One reason for the easy exchange of terms was that, with the help of mathematics, "in natural philosophy, truth is to be discovered by experiment and observation."[96] Praise by a humble contributor from York to the *Philosophical*

Transactions typified the trust placed in the experimental method: Experiments, he wrote, were the "sole method of becoming a truth."[97] By the end of the eighteenth century, experiments had "come so much into vogue, that nothing will pass, in philosophy, but what is either founded on experiments, or confirmed by them."[98] The significance of experiments for natural philosophy was far greater than the share of experimental papers in the *Philosophical Transactions*, only about one out of ten, might suggest.[99]

Having shown what the experimental philosophy could do, Newton explained what it was. The "Experimental Philosophy," he wrote, consists of two methods, "analysis" and "synthesis," in that order. Briefly stated, in the method of analysis, nature is analyzed by experiment, which separates the phenomena of interest from disturbing complications. Once the phenomena are known, by inductive reasoning, their regularity is generalized and stated as a law. The method of synthesis then takes over: From the law regarded as a principle, other phenomena are deduced, which are either known phenomena that were not considered when establishing the law, or new phenomena, the subject of further experiments. The methods of analysis and synthesis were ancient, but Newton's formulation of them as a unified method of experimentation and mathematics was received as new. Justified by Newton's authority, and by their success, the methods remained in standard use in research, and sometimes also in the organization of treatises,[100] throughout the eighteenth century.[101] Investigations of nature were expected to hold to the "*double test*" of the analytic and synthetic methods.[102]

Newton wrote, "The whole burden of philosophy seems to consist in this—from the phenomena of motions to investigate the forces of nature, and then from these forces to demonstrate the other phenomena."[103] We recognize in this statement the dual methods of the experimental philosophy reformulated in the language of forces. Later in this book, we consider theories that illustrate both formulations.

The methods of the experimental philosophy lent authority to the inquiry into nature, gave it its reason. Natural philosophers understandably valued their methods, but they did not credit them with every scientific advance. The author of a text on the experimental philosophy observed that discoveries were made not by "painful inductions," not by "investigation," but by "accidental experiments."[104] Trusting to their methods, natural philosophers at the same time accepted the unruliness of scientific practice.

Natural philosophy was identified with another philosophy, the Newtonian, commonly known as the "true philosophy." Except for being expected to last to the end of time, the Newtonian philosophy meant different things to different people. It was "as multivaried and as diverse as the interests of [Newton's] apostles and their world cared to make it," according to a scientific dictionary early in the century. Its more important meanings were given there only to be repeated in another scientific dictionary at the end of the century. One meaning was a new form of the "corpuscular philosophy," a kind of atomism which stood in opposition to the ancient and recent Cartesian forms of that philosophy; another meaning was Newton's discoveries; another was the principles and methods by which he made his discoveries. A related meaning was the "Mathematical and Mechanical Philosophy," which treated physical bodies mathematically and explained their phenomena by the laws of motion.[105]

The "mechanical philosophy" offered a unified, comprehensive explanation of the physical world; that was its beauty. Several versions of the philosophy existed in Newton's time,[106] although he did not refer to his work as such, and for the likely reason that he did not know the mechanical cause of forces.[107] The name is only infrequently encountered in British writings from the eighteenth century, but a common theme persisted: Nature is a machine, governed by laws binding matter, motion, and action of some kind.

The mechanical philosophy originally allowed only one kind of action, contact action, or impulse, since it was thought to be the only action that was not occult. This meaning of "mechanical" continued into the eighteenth century, with a following that included religious writers such as John Hutchinson, who accused Newtonians of promoting atheism.[108] In one version of the philosophy, effluvia, or invisible particles proceeding from attracting bodies, were assumed to act by impulse on passive, solid matter to produce the effects of attraction.[109] In another version, attraction was explained by aetherial particles moving in all directions with great speed.[110]

A restricted interpretation of "mechanical philosophy" was Newton's laws of motion or, yet more restricted, these laws as applied to visible motions, which leave the properties of bodies unchanged. The nineteenth century again widened the interpretation to make energy together with its heat equivalent the principal concept of physical science.[111] With his theory of heat, Cavendish pointed to this yet-distant energetic phase of the mechanical philosophy.

Method of Analogy

As a method of reasoning, analogy entered fundamentally into the work of the natural philosopher. Playfair characterized it as the inferring of like causes from like effects.[112] Thomas Young considered the statement that like causes produce like effects the "most general and most important law of nature," upon which "all analogical reasoning" is based.[113]

The power of analogical reasoning in natural philosophy derived from the analogy of nature, on which natural philosophers routinely quoted Newton: "[N]ature is very consonant and conformable to her self." For illustrations, they had to look no further than the *Principia*, where, for example, Newton developed laws of water waves from the familiar theory of the pendulum by likening the rise and fall of water in a U-tube to pendular vibrations.[114] The failure of a scientific explanation to be consonant with the analogy of nature counted as an a priori argument against it.[115]

The method had a wide following in the eighteenth century. Priestley said that every deliberate experimental discovery was made with the help of an analogy.[116] The author of a mechanics text said that every general fact or natural law was founded on analogy, on the comparison of the present with the past, ensuring the constancy of the phenomena.[117] To the author of a text on natural philosophy, the only acceptable proofs in natural philosophy were analogies, yielding probabilities, not certainties.[118] With the same confidence, the natural philosopher Richard Kirwan said that to explain a phenomenon, the method of "analogy, similarity or coincidence" was "by far the *most perfect and satisfactory*."[119]

In reasoning about things that cannot be experienced directly such as the behavior of matter at the level of particles, the method of analogy was indispensable, as a contributor to the *Philosophical Transactions* explained:

> In attempting to investigate matters too subtile for the cognizance of our senses, the only method, in which we can reasonably proceed, is by inferring from what we know in subjects of the same nature: and our conclusion thus inferred, concerning the subject sought, will be firmer and more unquestionable, in proportion as it resembles the subject known. But if the subjects be really of the same kind; if no difference can be shewn between them, in any respect material to the inquiry, in which we are engaged; in this case our inference from analogy becomes the very next thing to a physical certainty.[120]

Not everyone accepted inferences of this kind. The use of the heat of hammering as an argument for Newton's view of heat as the motion of particles showed, one investigator wrote, "how little trust ought to be paid to analogical reasonings in physical subjects."[121] Disagreeing with this objection, Cavendish built a Newtonian theory of heat on the analogy between the motions of particles and the phenomena of heat.

Mathematical Philosophy, Instrument of Theory

Theoretical physicists today regard mathematics as a "tool for reasoning" about nature, an "extremely useful tool";[122] further testimony on this obvious point is unnecessary. In his book on Newton's discoveries, Maclaurin said the same thing, that mathematics was the "instrument" by which Newton was able to accomplish his work. Maclaurin did not know if Newton showed more skill in "improving and perfecting the instrument, or in applying it to use."[123] As Newton needed the instrument, "sublime geometry," to write the *Principia*, so did those who would carry physics beyond that work. The lesson was not lost on Cavendish, for whom mathematics was the optimum instrument of "strict reasoning" in natural philosophy.

Up to the time of Newton, mathematics and physics were usually regarded as separate. Newton, too, made that distinction, but with the *Principia*, one commentator observes, he "bridged the gulf" between them.[124] His mathematics was the "skeleton of an explanatory scheme of such physical concepts as *power*, *propensity*, *cause*, *force*," his objective "a physics, and ultimately a natural philosophy."[125] In Newton's own words, in the *Principia* he cultivated "mathematics as far as it relates to philosophy." Lest readers find its mathematical principles "dry and barren," he included "scholia," discussions of experiments on physical bodies to which the mathematics was directed.[126]

Going by various names, "mixed mathematics," "physico-mathematical science," and "mathematical sciences," mathematical natural philosophy treated quantity as subsisting in physical bodies, in contrast to "pure mathematics," which treated quantity abstractly.[127] With his usual acumen, Newton foresaw the future of natural philosophy in those subjects he himself had not yet subjected to mathematics. Electricity was next in line, with heat not far behind; their experimental exploration occupied the middle part of the eighteenth century.

Theories of electricity and heat became exact before becoming mathematical. Benjamin Franklin's explanation of the Leyden jar, the electrical condenser, was not mathematical or even quantitative, but it was exact in that it provided unambiguous, verifiable predictions of the outcome of experiments.[128] That in itself was a considerable accomplishment, but as it turned out, his explanation was exact in the mathematical sense as well. His leading concept of an electric fluid proved capable of quantification and mathematical development, which was not overly long in coming. A student when Franklin's concept was introduced, Cavendish was forty when he published a masterful mathematical theory of the science of electricity based upon the hypothesis of a Franklin-like electric fluid. The study of heat had a similar history: twenty-five years separated Black's introduction of the concepts of specific and latent heats from Cavendish's mathematical theory of the science of heat. By the end of the century, in research in all departments of natural philosophy, it was understood that theoretical reasoning, quantitative experiment, and mathematics were to be joined as much as possible.

To bring mathematics to bear on subjects such as electricity and heat, three tasks had to be addressed. The first was to introduce concepts that made quantitative statements meaningful, for example, the concept of force. The second was to designate "mathematical measures"; the *Principia* began with measures of physical quantities, for example, measures of force.[129] The third was to express the results of experiments in numbers for comparison with the measures; this called for "mathematical" measuring instruments, from thermometers delicate enough to insert into the anus of vipers to machinery heavy enough to measure the force of cannon balls.[130]

Mathematical instruments made great advances in the eighteenth century, if by later standards they were still primitive. Those of reasonably high precision, the micrometer and the balance, measured length and weight. Length, the measure used by surveyors, carpenters, and tailors, was the natural philosophers' measure of velocity, the length of free fall required to generate a given velocity. It was their measure of other physical magnitudes as well: Small intervals of time were measured by lengths of pendulums; pressures were measured by lengths of columns of mercury or other liquids; temperatures were measured by lengths of the same kind. Weight, the measure used by merchants, manufacturers, and chemists, was the natural philosophers' measure of gravitational attraction and mass. Normally, the acceleration of gravity, an "accelerating force," was taken to be unity, and other accelerations were expressed in relation to it as pure numbers.[131]

Natural philosophers extended measures from mechanics to heat, electricity, and other subjects. Heat could only be compared with itself, but in "Heat" Cavendish gave to sensible and latent heats an additional, mechanical measure, work. Electricity, too, could be compared only with itself, but as a moving force, it could be measured by forces appearing in mechanics, for example, gravitation, as Cavendish measured it in "Heat." "This observation is momentous," the natural philosopher John Robison wrote, for the mechanical philosophy then becomes one of the "*disciplinae accuratae*."[132]

Valued for its strict reasoning, mathematics enabled natural philosophers to trace physical relationships through a "long Process of Reasoning, and with a Perspicuity and Accuracy which we in vain expect in Subjects not capable of Mensuration."[133] To this end, they called upon both of the great branches of mathematics, algebra and geometry. In algebra, symbols could stand either for abstract numbers or for physical quantities such as force and brightness of images, and physical quantities could also be represented by lines and other figures in geometry. The Professor of Astronomy and Experimental Philosophy in Cambridge described lines appearing in his optical drawings as "physical lines," consisting of "physical points."[134] It worked the other way around, the same author writing, "I suppose the chord [musical string] to be uniform and very slender, or rather to be a mathematical line."[135] In principle, natural philosophers could express their physical quantities as symbols or as lines indifferently, since they could move readily between the two representations, using algebra to solve geometrical problems, and conversely. "The mutual intercourse of these two sciences has produced many extensive and beautiful theories," Maclaurin wrote,[136] and natural philosophy was the richer for it.

Reasoning in natural philosophy led to probability, and in mathematics, to certainty; but aside from that basic difference, the methods, concepts, and formulations of mathematics and natural philosophy had multiple points in common—the two bodies of learning mirroring one another, as the following observations illustrate. First, in principle, they excluded the same things, for example, metaphysical speculations. Mathematics inquired "into the relations of things rather than their inward essences," an advantage, because ideas of relations of things were clearer than those of essences of things. Natural philosophy likewise investigated relations, not essences, and achieved clarity into the bargain. Calling upon the same logical operations, mathematics and natural philosophy conveyed the same rewards and incitements. Both formed the habit of "thinking *closely*, and reasoning *accurately*."[137] Both satisfied a "*natural* desire" for knowledge, and conferred "great satisfaction and delight . . . in the discovery and possession of truth." Both met spiritual needs, inducing a regard for the "infinite knowledge and wisdom, power and goodness of the *Almighty* Creator."[138] Both were "sublime" and "noble," possessed "beauty," and had the same principal author, the "divine" Newton.[139]

Mathematics and natural philosophy used the same terms. Both were expressed in quantities, their most obvious common denominator, the source of their constant, fruitful interchange. Both spoke of "theory and practice," of "methods" including those of inductive "analysis" and deductive "synthesis," of "rules" together with the goal of making the rules as simple and general as possible, and of "principles," the fewer the more scientific. Mathematics and natural philosophy used the same categories of strict reasoning: "axioms," or irreducible truths; "postulates," or statements of the possibility of doing certain things; "problems," or things proposed to be done; "theorems," or demonstrated truths; and "propositions," or theorems or problems. These conscious outward resemblances of the mathematical and natural philosophies rested on what was thought to be their inner affinity, a profound truth about nature: Mathematics was the "language of nature."[140]

The Newtonian calculus, which "discovers and opens to us the Secrets and Recesses of Nature," illustrates the shared language of natural philosophy and mathematics. Founded on the principle that a quantity is generated in the way that a line is generated by motion, this calculus referred to the "power" to generate a quantity as a "velocity," to the generated quantity as "flowing," and to the velocity of its flow as its "fluxion." The velocity was meant not literally, but as an analogy, a mechanical image of the abstract idea of a fluxion, the rate of increase of any generated quantity. It was also directly useful in describing the motion of bodies in the real world, and because fluxions existed for any continuously variable quantities, in principle fluxions described not only motions but any kind of phenomena encountered by the natural philosopher, for example, those of electricity and heat.[141]

Newton having demonstrated once and for all that aspects of nature can be described mathematically, and with extraordinary precision, his successors might be expected to have learned mathematics as a matter of course. They often did not, however, as the article "Physics" in the *Encyclopaedia Britannica* from 1797 ruefully noted:

> A notion has of late gained ground, that a man may become a natural philosopher without mathematical knowledge; but this is entertained by none who have any mathematics themselves; and surely those who are ignorant of mathematics should not be sustained as judges in this matter. We need only appeal to fact. It is only in those parts of natural philosophy which have been mathematically treated, that the investigations have been carried on with certainty, success, and utility. Without this guide, we must

> expect nothing but a school-boy's knowledge, resembling that of the man who takes up his religious creed on the authority of his priest, and can neither give a reason for what he imagines that he believes, nor apply it to any valuable purpose in life.[142]

Natural philosophers who lacked mathematics could understand Newton's *Opticks* but not his *Principia*. They necessarily limited themselves to the more experimental parts of natural philosophy, which in any case temporarily provided them with sufficient opportunity. As Newton did in *Opticks*, they wrote their papers and books with "Proofs by Experiment" rather than demonstrations.

The experimental natural philosopher Benjamin Wilson addressed a book on electricity to a colleague, who earlier had applied his "mathematical abilities" to Wilson's work. Wilson encouraged him to continue in that vein, to "treat this part of philosophy in the same manner as Sir Isaac Newton has done the great subjects in his *Principia*."[143] Wilson did what he himself could do in this direction by giving a mathematical discussion of the acceleration of electrical particles in a discharge cylinder, and presenting his experiments in the axiomatic Euclidean format of the *Principia*, complete with a General Scholium. His mathematical aspirations for electricity belonged to the same decade as Cavendish's mathematical theory of heat.

Other natural philosophers learned their mathematics thoroughly. As instructional guides, they had Newton's geometrical method from the *Principia*, and his posthumously published lectures on the method of fluxions, as well as a good number of other sources through the century, some of the best of which originated in heated controversies of the day. The most important of these, Maclaurin's *Treatise on Fluxions* in 1742, was written to dispel doubts about Newton's method. It retained Newton's notation for a fluxion, a marvel of concision, a dot over a variable, $\dot{x}$, indicating the instantaneous change of the variable, relative to time. Regarded as the first logical, systematic presentation of the Newtonian form of the calculus, Maclaurin's treatise was a major contribution to the "mathematical philosophy."[144] It and other books on fluxions contained solutions to physical problems and might include discussions of the match between solutions and observed phenomena, in this way disseminating a knowledge of fluxions and of the mathematical parts of natural philosophy at the same time.[145] If in some instances prematurely,[146] to varying degrees, all scientific studies were subjected to mathematical treatment in the eighteenth century. In Cavendish's time, about a fifth of the papers appearing in the *Phil-*

osophical Transactions belonged to optics, pneumatics, and the other parts of mixed mathematics.[147]

Let us see what a mathematically competent British natural philosopher knew. He was familiar with the concept of a "function," a quantity dependent on one or more variables, a starting point of mathematical reasoning throughout natural philosophy.[148] He was knowledgeable in the elementary branches of mathematics. The author of a textbook on the fluxional calculus assumed, perhaps optimistically, that his readers already knew "perfectly" arithmetic, geometry, algebra, doctrine of proportions, logarithms, trigonometry, and mechanics.[149] He knew the calculus, which included the method of infinite series. Although it was relatively easy to find the fluxion, or derivative, of a variable quantity, it was harder to proceed in the inverse direction, to find the original variable quantity, or integral, of a given fluxion of any complexity. Infinite series were the recourse: A given fluxion was expressed as an infinite series, the simple terms of which were then readily integrated one by one, and the first few integrals were kept as an approximation to the answer. "Laborious and disagreeable" as the method of infinite series could be in practice,[150] it was indispensable in dealing with physical problems, and natural philosophers like Cavendish were thoroughly at home with it. Beyond the rudiments of the calculus and the method of infinite series, the mathematically knowledgeable natural philosopher was familiar with bodies of mathematics arising jointly from the calculus and problems in gravitational astronomy, pendulum motion, elasticity, fluid flow, propagation of sound, and other physical topics. These included ordinary differential equations, partial differential equations in the case of functions of several variables, and the calculus of variations for finding maximum and minimum values of variable quantities. By the time Cavendish entered his career, these subjects had begun to be studied in their own right, as parts of mathematics separate from physics and the calculus. The complete natural philosopher probably also knew something about other mathematical subjects having as yet little or no use in his work, such as probability, differential geometry, and number theory. Cavendish studied them all.

The observation has been made that before the nineteenth century, the only parts of physics requiring advanced mathematical skill were mechanics and hydrodynamics.[151] That is correct if it is recognized that a nonmechanical part of physics was developed as a mechanical theory. Cavendish's electrical theory was, in part, a theory of fluid mechanics, and in working it out mathematically, he did what Newton had done in fluid mechanics: He relied on special cases, idealizations, approximations, and guesswork. Acknowledging the mathematical difficulties of his theory, Cavendish wrote: "I am obliged to make use of a less accurate kind of reasoning"; it "does not appear for certain"; "it is likely," "very nearly," "cannot tell." He hoped that "some more skilful mathematician" than he would prove certain of his propositions.[152] By 1771, electrical theory already posed mathematical complexities beyond the considerable ingenuity of Cavendish, whose mathematical knowledge and skill were advanced for the time.

The mathematics used by British natural philosophers was distinctive. In the wake of a dispute between Newton and Leibniz over the invention of the calculus, the British took up Newton's form of the calculus, fluxions, rather than Leibniz's analytic form.[153] They also preferred geometry to algebra. Newton had written the *Principia* using geometrical methods rather than fluxions, although the latter, too, was described as a "new geometry," unknown to the ancients.[154]

Thomas Simpson, an exception among British mathematicians, did not share in the tendency to dislike everything "performed by means of *symbols* and an *algebraic* process." Neither, he said, had Newton, who used the algebraic method in the treatment of bodies moving through resisting media, and elsewhere. The alternative, the geometric, or "synthetic," method of demonstration, was not always best, a truth that British mathematicians had avoided at their cost. By cultivating the algebraic method, Simpson correctly observed, "Foreign Mathematicians have, of late, been able to push their Researches farther, in many particulars, than Sir Isaac Newton and his Followers here, have."[155] British mathematics eventually underwent needed reform, but that came only in the nineteenth century.

Even as they elected not to use it, British investigators were exposed to the Leibnizian calculus, which turned up occasionally in foreign papers in the *Philosophical Transactions*.[156] Leibniz's notation, in which the differential dx stood for an infinitesimal increment of the variable x, was powerful, and for the calculus of variations nearly indispensable. Perhaps for this reason, and definitely because so many able mathematicians used the Leibnizian calculus, Maclaurin took pains to demonstrate a "harmony" between the method of infinitesimals and Newton's method of limits,[157] but otherwise he led the autonomous development of British mathematics.

By the time Cavendish studied higher mathematics, the Newtonian school of mathematics and mathematical physics was well established in Britain.[158] Maclaurin's book was several years old, and the polemics over the invention of the calculus and the method of fluxions had subsided, the passions had quieted, and the concept of the limit of a ratio of vanishing quantities had been clarified.

Cavendish could concentrate on mastering the methods of this recent branch of mathematics of proven usefulness in natural philosophy. He learned British mathematics, and that is what he used.

There are scientists today who think that the universe is a structure precisely governed by "timeless mathematical laws," that the deeper we probe, the more the physical world dissolves into mathematics,[159] and that the "complete merging of theoretical physics with pure mathematics" is a realistic prospect.[160] Sturdy empiricists that they were, eighteenth-century British natural philosophers did not talk that way, but in their treatises on mechanics and their colleagues' treatises on fluxions, they came close.

Theories

Questions, Conjectures, Hints

Natural philosophers expressed their convictions with various degrees of forcefulness: great confidence, bordering on mathematical certainty; prudent open-mindedness, inviting criticism; and tentativeness, often in the form of questions.

Newton had a just appreciation of the power of the question. "This brilliant genius," Einstein wrote of Newton, "who determined the course of western thought, research, and practice like no one else before or since," was led to his greatest achievement by a "question": Is there a simple rule by which the motions of the planets can be calculated?[1] His answer to this question was, as we know, a book, the *Principia*. Like Einstein, drawn to riddles, Newton viewed the universe as a hard riddle indeed. He did not solve it completely, but he gave it a clear formulation, and he turned the riddle into a productive genre of scientific exchange, "Queries." Through his queries, he left challenges and hints for his successors, backed by his infectious curiosity about the universe and his optimism about science.

Newton's queries appeared as an appendix to his *Opticks*. Although he had intended to make more experiments on light, by the time he published his treatise, his experimental days were behind him, and instead of giving experimental answers he posed questions, hoping to prompt "a farther search to be made by others."[2] His first queries dealt appropriately with optics, to which, in keeping with his questioning way, in subsequent editions he added, new queries dealing with other branches of science, the last, Query 31, containing the outline of a complete natural philosophy. Newton's younger colleague and editor Roger Cotes observed that "whoever will read those few pages [the last query] of that excellent book, may find there in my opinion, more solid foundations for the advancement of natural philosophy, than in all the volumes that have hitherto been published upon that subject."[3]

Cotes's admiration was widely shared. Playfair said that Newton was "hardly less distinguished from others by his doubts and conjectures, than by his most rigorous and profound investigations."[4] Newton's readers took his queries to be a major legacy, the most valuable part of *Opticks*, a blueprint for the future of natural philosophy, and proof of the author's love of wisdom and rejection of dogmatism.[5] In their writings on natural philosophy, they not only frequently cited Newton's queries, but also proposed queries of their own, or their affirmative counterparts, conjectures and hints. Like Newton, they relegated their queries to the end of their writings, after the solid facts, with the goal of stimulating scientific research. An avid student of Newton's queries, Stephen Hales suggested a reason why his colleagues welcomed the open-ended scientific question. Without tentative assertions, science would advance but slowly, "for new experiments and discoveries usually owe their first rise only to lucky guesses and probable conjectures."[6]

Queries served as a tentative kind of scientific theorizing. Newton put forward his theory of heat most fully in queries, presumably Cavendish's source. The author of a paper in the *Philosophical Transactions* on the colors of the flame of burning substances, which he attributed to different degrees of attraction between the particles of light of different colors and the particles of substances, concluded his discussion with a "theory," advanced as a loose series of queries about weaker and stronger attractions.[7] Of the "excellent theory" of positive and negative electricity, as Priestley called it, its originator William Watson spoke more cautiously, referring to his understanding of electricity as a "system," but not quite. What he proposed was not a "System itself," but a "rude Outline of a System," and he made clear his tentativeness by stating his theory in the form of "queries" at the end of his paper.[8]

Despite their form, Newton's queries were sometimes taken as statements of his considered views, as they perhaps in part were intended. That also happened to authors who lacked Newton's authority, possibly with unwanted consequences, as the experience of Alexander Wilson illustrates. Professor of Practical Astronomy in the University of Glasgow, Wilson turned his telescope to a giant sunspot, closely observing its central, dark nucleus and the penumbra surrounding it, and noting how appearances changed from day to day as the spot moved across the sun's disk. By geometrical reasoning, he concluded that this and also other spots are excavations in the luminous matter of the sun. That much constituted part 1 of Wilson's paper. Part 2 contained his explanation, which he presented as "queries," "conjectures," and "hints." Although he was in agreement with Newton that the only way to truth was by induction from experiment, he also agreed with Newton that conjectures have a place in natural philosophy, on occasion leading to a correct understanding of the phenomena. Having given in part 1 an "experimental" proof that sunspots are excavations, in part 2 he conjectured that the sun is a cold, dark body surrounded by a luminous fluid of foglike consistency, through which an elastic fluid rises, creating excavations. He proposed an "experiment" with a telescope to confirm his conjectures.[9]

Nine years later Wilson published a second paper on sunspots. In the meantime, the French astronomer J. J. L. de Lalande had criticized his explanation, particularly his manner of explanation, which he found loose and problematic. This was in part an instance of Anglo-French incomprehension, but implicit in the French criticism was a valid point: Queries could be used to evade responsibility for a mistaken hypothesis, or to claim credit for a correct one. As a rejoinder, Wilson recalled the organization of his earlier paper, an inductive part followed by a conjectural one. Lalande had overlooked Wilson's careful distinction between "fact and any thing like to theory." He had propounded his "theory" in the "form of queries" because he wanted to arouse curiosity and excite observation by others without misleading them. With this clarification, he sought to rescue the inductive part of his paper from "being drawn into the eddy of some treacherous theory." Lalande had a different notion of the nature of sunspots, thinking them rocklike outcroppings, and although he presented this notion as a mere supposition, he was not spared. Wilson referred to Lalande's explanation of sunspots not as a supposition but as Lalande's "theory," which he set out to destroy. Lalande's criticism led Wilson to acknowledge that, despite their form, his queries could be taken as a seriously proposed theory and, in response, to dissociate his work on sunspots from the "sandy foundations" of all theory whatsoever. He was confident of his observations, but not of his explanation, conscious of his and others' necessarily imperfect knowledge of the "vast range of physical causes which obtain in the universe" and its "numberless unthought-of energies."[10]

A query might be answered in the spirit in which it was posed, tentatively. Our example is again provided by Alexander Wilson, who addressed a question that Newton had raised in the final query of *Opticks*: What prevents the fixed Stars from falling upon one another owing to their mutual attraction? He gave his answer, projectile and periodic motions, not as facts but as hints, which he hoped would lead to observations confirming or denying them.[11] In natural philosophy, one question led to another, a perpetual motion of thought.

If an investigator presented facts without any kind of explanation, his publication would likely have seemed incomplete. A query, conjecture, or hint was evidence that the author was not a brute compiler of information, and it acknowledged that the reader was an intelligent being wanting enlightenment and stimulation. A theoretical viewpoint was usually present. At the end of a paper announcing the discovery of a periodicity in the brightness of the star Algol, the author said that his purpose was to present facts, and that it was too early to make a "conjecture" about the cause of the variation, but he made one anyway.[12] Two years later another astronomer referred to the same Algol "conjecture" as a "hypothesis."[13] The term "conjecture" was used interchangeably with "hypothesis," our next topic.

Hypotheses

Hypotheses were a concern of British empiricism from the start. Locke accepted them, if guardedly, as a guide in the search for the truths of nature, which he believed extended to truths about the realm of invisible particles. He held two criteria for deciding between hypotheses—greater explanatory reach, and freedom from inconsistency and absurdity—which were met, he believed, by the hypothesis of heat as the motion of the invisible particles of bodies. Cavendish agreed, and for much the same reasons as Locke.[14]

Today science and the hypothetical method are practically synonymous, but that was far from the case in the eighteenth century, as was evident from the standard British version of the history of the Scientific Revolution: Hypotheses had passed for science until Francis Bacon introduced the right method, whereupon by applying the rules of experiment and observation, in a short time, natural philosophy advanced more than it had in all previous ages, and with the explanation of the true system of the world "all physical hypotheses vanished, like phantoms before the philosophy of Newton."[15] A mid-century book on Newton's achievements labeled earlier thinking in science "extravagant fictions," characterized by fondness for the marvellous, metaphysics, complete systems deduced from first causes, and "hypotheses."[16] This belittlement of the past was an echo of Newton's battles. Late in life Newton said that Descartes was the reason he had written the *Principia*, and from the vantage of his followers, Newton's rejection of Cartesian vortices permanently tarred hypotheses. The lesson from history was that experiment and observation led to progress, and hypothesis, to stagnation, a reputation hard to live down. Hypotheses were the unacceptable face of natural philosophy.

"Hypotheses" were statements that could not be deduced from phenomena, according to Newton, who ruled that they had no place in the experimental philosophy.[17] As good as his word, the opening sentence of the primer of the experimental philosophy, *Opticks*, reads: "My Design in this Book is not to explain the Properties of Light by Hypotheses, but to propose and prove them by Reason and Experiments."[18] In the first part of *Opticks* and the last part of the *Principia*, he deduced conclusions from the phenomena in conformity with his understanding of the experimental philosophy, separating hypotheses as much as possible from the experimental philosophy, and relegating them to queries and scholia.[19] Newton's ruling and example were cited with approval through the eighteenth century.

Generally speaking, when a natural philosopher offered the scientific world a hypothesis, he risked criticism. An author rebuked the inventor of an electrical theory for proceeding from a hypothesis: had he begun with facts, and from the facts deduced the hypothesis as a general conclusion, he would have been "more philosophical."[20] Having conjectured on the nature of meteors, an author reined in his imagination, observing that "one fact in philosophy well ascertained is to be valued more than whole volumes of speculative hypotheses."[21] Another wrote that "theories formed on mere hypothesis are always uncertain, and little to be depended upon."[22] William Herschel called the proper motion of the sun "my hypothesis," but it was not a "mere hypothesis," because in the case of several stars proper motion was an established fact, and the sun was a star.[23] When a genuine discovery such as the law of gravitation was referred to as a "hypothesis," it was regarded as a misuse of language, a habit of speech befitting an earlier era of science.[24]

Yet even Newton's warmest disciples recognized that the master used a hypothesis now and again when it suited his purpose. An example of an unexceptional use was his reasoning about the figure of the earth: from the hypothesis that the axis and equatorial diameter of the earth are as 100 to 101, by proportional reasoning he determined that the real ratio is 229 to 230.[25] For his strictures on hypotheses, Newton could be faulted. It was pointed out that he had condemned hypotheses in the queries of *Opticks*, the very part of the book that was "intirely hypothetical." Moreover, he had spoken there positively about four causes of refraction, at least three of which had to be wrong.[26] The ways in which Newton used the term "hypothesis" have been counted; the total is not small.[27]

If on occasion natural philosophers spoke disparagingly of hypotheses, they used them—cautiously, it is true, but unapologetically. Priestley, who regularly acknowledged hypotheses as the starting point of his experiments, advocated their general use as an aid in discovery while cautioning against forming too great an attachment to them.[28] Others held similar opinions. Disagreeing with Newton on the explanation of comets' tails, the Professor of Natural Philosophy in Dublin, Hugh Hamilton, offered a hypothesis of his own together with a sober recommendation of hypotheses in general: "I think that Conjectures, or Hypotheses, when rendered probable by some Experiments, and proposed with Caution, may be of great Use by directing our Enquiries into some certain Channel."[29] The Jacksonian Professor of Natural Philosophy in Cambridge, Isaac Milner, advised his students that hypotheses were "dangerous" if taken too seriously, but that they could help investigators "devise new experiments for some definite purpose."[30] Robison, although unwilling to accept hypotheses as causal explanations, valued them as "conjectures serving to direct our line of experiments."[31] The author of a paper on the ascent of vapor defended his "hypothesis" concerning a certain kind of matter, a fiery fluid, against Newton's criticism. He did not defend the hypothetical fluid, but rather the use of hypotheses in general.[32] Hypotheses allowed investigators to look beyond the evidence of their experiments into the twilit borders of "terra incognita,"[33] where new seas and continents lay, awaiting discovery.

Hypotheses proved indispensable in certain kinds of investigations, for example, in Cavendish's theories of electricity and heat. Inductive reasoning from the phenomena could not have led him to a physical cause as complex as the kinds of matter and forces with which he explained electrical phenomena, nor could it have led him to the mechanical concept by which he explained the phenomena of heat.

In researches falling outside the reach of experiment, natural philosophers might have no choice but to incorporate hypotheses into their explanations. John Pringle, who coordinated a wide-ranging investigation of a certain fiery meteor, rejected the hypothesis that fiery meteors are sulfureous vapors rising from the earth, or that they are lightning. The hypothesis he favored was that they are bodies independent of the earth, perhaps orbiting about a center.[34] In printed directions to observers of fiery meteors, the Astronomer Royal Nevil Maskelyne said nothing about their nature, but in an accompanying letter, he referred to Pringle's hypothesis as plausible. Charles Blagden, Secretary of the Royal Society, in conjunction with queries to observers, rejected Pringle's hypothesis that fiery meteors are "terrestial comets" that excite light when they enter the atmosphere, proposing instead that they are masses of electric fluid attracted to or repelled by the earth's magnetic poles.[35] Pringle's, Maskelyne's, and Blagden's purposes in mentioning hypotheses were the same: to explain, however tentatively, and to encourage observers to make reliable reports of a sporadic, transient, and puzzling phenomenon.

A scientific dictionary characterized an hypothesis as a "system laid down from our own imagination," proposed as a possible cause of phenomena.[36] In their search for truths of nature, natural philosophers acknowledged a widening activity of the mind, warily entertaining hypotheses as the offspring of legitimate scientific questions instead of reflexively denouncing them as the progeny of fantasy and license. Striking a balance between belief and skepticism, they were willing to assume a statement without assuming its truth. They grew accustomed to working with hypotheses, to granting them a role in directing inquiry, while retaining experiment and observation as the arbiter of truth, the essential point of the experimental philosophy. Their admission of hypotheses into natural philosophy expanded the scope of theory, and at the same time it introduced a new complexity into reasoning about nature, which proved permanent, as today's philosophers of science can attest.

The compiler of a scientific dictionary at the end of the eighteenth century observed that the latest, best authors no longer used hypotheses, but obeyed Newton's injunction against them, and reasoned only from experience.[37] That was not to be the last word on the subject. Had natural philosophers owned a crystal ball, they would have gazed with wonder upon what followed their acceptance of hypotheses two hundred years later. According to certain respectable positions in the philosophy of science today, science follows no method at all, but proceeds from no-holds-barred, freely invented hypotheses.[38]

Theories

Erasmus Darwin chastised colleagues who disliked all theory, who forgot that "to think is to theorize."[39] Theories fulfill a need to know more than what is presented in experience, the natural philosopher and chemist William Nicholson said: Every "theoretic system" is based on the mind's inclination to make general inferences from particular facts.[40] The mind, it seems, is a born maker of theories.

In this activity, natural philosophy offered the mind ample opportunity, as Newton's case illustrates. Unrivaled in its explanatory power, his theory of gravity gave rise to numerous gravitational theories, starting with his own: his "lunar theory," "theory of the comets," and theory of the precession of the equinoxes, to mention only a few.[41] Eighteenth-century natural philosophers had before them abundant examples of theory originating with the highest authority.

On the making of theories, let us turn to a later authority. The physical theorist, according to Einstein, has two tasks: first, to establish principles, and second, to deduce theoretical consequences from the principles. The latter requires a knowledge of mathematics, which can be acquired by anyone with the requisite aptitude and industry. The first task is different, more difficult, not found in books. Having no method to guide him, the theorist grapples with the facts, seeking the appropriate generalizations to use as principles of reasoning. Knowing that without principles, facts are useless to him and that with them, he can reveal unforeseen relations between facts, the theorist's supreme challenge is the formulation of principles.[42] Eighteenth-century natural philosophers would have recognized Einstein's second half of the theorist's work as their synthetic method. The first half would definitely have intrigued them, Cavendish certainly. It resembled their own experience in arriving at provisional explanations, at hypotheses, an art for which they had neither method nor name.

Physics today has two main kinds of theory, one of which, "empirical" or "phenomenological," identifies regularities without explaining them, and the other explains the regularities.[43] These kinds we recognize as descendants of the two main tasks of natural philosophy, to "describe" phenomena and to "explain" them from causes.[44] There were two corresponding kinds of theory in our period. Newton's resolution of white light into colored rays led to a theory of the first kind, one based on experiment. This, the "true theory" of light, Priestley said, was the "model for all future inquiries into the powers of nature."[45] Newton's explanation of the phenomena of heat from the cause of heat, the mechanical vibrations of particles, was an example of the second kind. Cavendish made theories of heat of both kinds, one independent of any hypothesis, the other, our principal example in this book, based on Newton's hypothesis of the cause of heat.

A scientific dictionary used "theory" and "hypothesis" interchangeably;[46] an electrician called Benjamin Franklin's explanation of the Leyden jar a "hypothesis" and a "theory";[47] an astronomer called his explanation of the structure of the universe a "hypothesis" and a "theory."[48] Other authors wrote about hypotheses and theories as distinct. The philosopher Dugald Stewart observed that because hypotheses were "commonly confounded with theory," too much was demanded of them, resulting in a bias against them. Hypotheses, he explained, were "necessary for establishing a just theory," but they were only the first anticipations of the principles of a theory, a link in the chain of reasoning in natural philosophy. Without hypotheses, the mind could not conceive of the experiments necessary to establish the principles in the first place, but once the principles were extracted from the facts, they could be used to explain new phenomena, becoming then "legitimate theories."[49] The natural philosopher Tiberius Cavallo regarded hypotheses and theories similarly: Rival hypotheses led to experiments, which decided among them, the best hypothesis then becoming the basis of a "theory, *viz*, the real cause" of the phenomena.[50] Describing how hypotheses and theories worked together, Priestley said that a hypothesis, which was "nothing more than a preconceived idea of an event," might lead to further experiments, from which new facts arose, which in turn served to "correct the hypothesis which gave occasion to them." The theory, now corrected, led to more new facts, which brought the "theory still nearer to the truth." Eventually, by trial and error, feedback, and revision, the hypothetical method produced the desired result, all of the facts together with a "perfect theory."[51]

A theory carried more conviction than did a hypothesis. After explaining the aurora borealis, Cavendish said, "I wish it to be understood, however, that I do not offer this as a theory of which I am convinced; but only as an hypothesis which has some probability in it."[52] Playfair said that theories were confirmed by facts known independently of the phenomena they explained, whereas hypotheses were facts assumed to explain phenomena and had no other evidence of their reality.[53] Robison said that to explain phenomena, the best course was to make experiments and observations to discover the basic facts, then to establish general physical laws, and then to provide a theory showing how all the subordinate phenomena fell under them. To make a theory of this sort, he said, no hypotheses were required. If, however, the phenomena could not be so simplified, "we cannot establish those general laws which would be the foundation of a physical theory." We then had to resort to the next best method, which was to formulate a hypothesis suggested by the phenomena, and from the hypothesis together with the principles of mechanics to deduce a range of interesting consequences, a selection of which was then compared with observations. If the two disagreed, the hypothesis was rejected, but if they agreed, the hypothesis was admitted as probable. "We may then discover by this means parts of a hypothesis which must be admitted as true, although the hypothesis cannot be demonstrated in its full extent."[54] The first method, the original method of the experimental philosophy, led to a true physical theory. The second, the hypothetical method, led to propositions that were true but also to unconfirmed propositions, and for that reason it inspired less trust and more science. To natural philosophers, theory, hypothesis, and experiment were integral parts of a single process: scientific research.

Like authors of hypotheses, authors of theories could anticipate criticism. Even Newton had been criticized for several of his "Theorys,"[55] an encouragement of sorts to later theory-makers. Theories were inherently chancy; for example, they were often mistaken; they misled investigators;[56] they distorted the language of factual reporting;[57] they biased scientific judgment. Generally conceded, hazards like these were not seen as a persuasive argument for banning theories from natural philosophy. That had been tried in the case of hypotheses, and it had failed. The absence of theory was not a guarantee of scientific candor; as one natural philosopher observed, disbelief in theory was itself a theory.[58] The goal was not to shun theories, but to make better ones.

The mathematician and natural philosopher Samuel Horsley summarized the main uses of theory in natural philosophy:

> I am well aware, how little theory is to be trusted, in its *remote* conclusions, on account of the necessary deficiencies of the physical *data*, upon which its reasonings are founded. The true uses of it are, either to explain the mutual connexions and the dependencies of things already known, or to suggest *conjectures* concerning what is unknown, to be tried by future experiment. And he who applies it, with due circumspection, to these purposes, will always find it a useful engine.[59]

Few natural philosophers would have quibbled with Horsley's conclusion that theory is a "useful engine," or with the reasons he stated. Theories provided "approximations to the truth," which served until the "real laws of nature" were discovered.[60]

To be accepted, a theory had to agree with experiment, of course, although a theory could disagree and still be correct. Theories told what happens under given conditions or circumstances, Priestley said, and a perfected theory defined the circumstances of every appearance of the phenomena under consideration.[61] Upon reading Newton's discussion of fluids in the *Principia*, and subsequently carrying out experiments on the times of descent in water of hollow brass globes filled with various substances, Atwood reflected on the relationship of theory and experiment. Recognizing that mathematical exactness in experiments of this kind was unobtainable, he strove for attainable precision, which meant taking care to observe the conditions of the theory in the experiment. He attributed the differences between the times of descent calculated from Newton's theory and the times he observed to the difficulty of achieving the conditions rather than to a flaw in the theory.[62]

If the conditions of the theory were met, and the theory still disagreed with experiment, its truth was in doubt. Nevertheless, it was not summarily dismissed, because the error could as easily have arisen from experiment as from theory. It worked both ways:

> Every theoretical inquiry, whose basis does not rest upon experiments, is at once exploded in this well-thinking age; . . . But experiments themselves . . . are not exempted from fallacy. A strong inventive facility, a fine mechanic hand, a clear unbiased judgment, are at once required for the contrivance, conduct, and application, of experiments; and even where these are joined (such is the condition of humanity!) error too frequently intrudes herself, and spoils the work.[63]

Although as the generalization of "broad experience," experiment was the "supreme authority," a theory was not abandoned because of a solitary "anomalous" fact.[64] Experimenters fully accepted their fallibility, a self-criticism built into the experimental philosophy.

Bound together by logic, laws, concepts, and analogies, theories had also to agree with one another. Theories came with different levels of generality, the most general being theories of knowledge, variants of British empiricism, implicitly assumed. Next came theories of general methodology: The method of the experimental philosophy constituted the "entire theory of natural philosophy."[65] Next came theories of matter and force applying to all branches of natural philosophy: Boscovich's theory of universal force was an example. Next came theories of mechanics such as Newtonian mechanics; next came theories of other parts of natural philosophy such as Cavendish's theory of heat; next came partial theories such as Joseph Black's theory of latent heat; finally, there were theories of individual instruments, specific methods, and experiments.

The latter theories, the most specific, call for a brief clarification. The size, proportion of parts, and particular ways of making scientific instruments could "be known only by experience," and ignoring that reality, writers on the subject often lost themselves in "reveries,"[66] but the design, use, and explanation of instruments were guided by theories. Atwood took a "theoretical view" of a watch, comparing its motion with the laws of mechanics.[67] The electrical machine had an electrical "theory,"[68] Hadley's quadrant had an optical "theory,"[69] and so on through the parts of natural philosophy. Methods of inquiry had their own theories, for example, Herschel's method of the annual parallax of double stars.[70] Experiments, too, had theories of their own; for example, Jan Ingen-Housz presented his experiments on photosynthesis together with "theory of exp. I," "theory of exp. II," and the like.[71]

We close with a word about a standard distinction of the time, "theory" versus "practice." Philosophy was divided into two branches, "theoretical" and "practical"; the first branch was contemplative, the second active.[72] Astronomy was likewise divided into "theoretical" and "practical," or observational: We read that the instruments of astronomy raised difficult problems, requiring the "most accurate theoretical investigation, and the utmost refinement of practice."[73] Machinery, manufactures, and other human productions had their "theory" and "practice": We read that no propositions in rational mechanics or experimental chemistry were so "purely theoretical as to be totally incapable of being applied to practical purposes."[74] We may question the practical benefits of eighteenth-century theories, but they undoubtedly provided concepts and laws useful for discussing, in a general way, machines and other technology.[75] In this sense, theories provided the artificial world with what they provided the natural—understanding; theories illuminated the reason in machinery as they did the reason in nature. As the century progressed, the Royal Society and its journal gave less attention than it had earlier to reports on practical applications, for which there were then other outlets.[76] Theory as distinguished from practice is not our main concern in this section; we return briefly to the subject when discussing Cavendish's theory of heat.

Fictions

"Fiction," a word with many uses in the eighteenth century, comes from "fingere," to shape or to mold. The first definition, from 1784, in the *Oxford Universal Dictionary* reads, "the action or product of fashioning or imitating."[77] "Fiction," in that sense, can be applied to theory in natural philosophy, although to oppose observation to theory and to denigrate all theory as "mere fiction" were judged harmful to science.[78]

Since in "this well-thinking age," the eighteenth century, "every theoretical activity" rested on experimental facts,[79] we have to remind ourselves that fiction was valued, too. The novel was not newly invented then, but the "belief that fiction was artistically and intellectually worthy of a major talent" was new to British letters.[80] In his commentaries on English law, which abounded in legal fictions, *fictione juris*, the jurist William Blackstone repeatedly instructed his readers in their usefulness.[81] The philosopher David Hume said that the supposition that things in the world persisted when he was not perceiving them was a fiction, and also that there was hardly a moment in his life when he did not live by that fiction.[82] The logician of language and author of *Theory of Fictions*, Jeremy Bentham wrote that without "fiction, the language of *man* could not have risen above the language of *brutes*."[83]

Such appreciation notwithstanding, fiction had a checkered reputation. Unconstrained by fact, it lent itself to devious uses. If, as the logician claimed, fiction was the "coin of necessity" to humans as reasoning creatures, in the hands of priests and lawyers "it has had for its object or effect, or both, to deceive," and in those of poets "to amuse, unless it be in some cases to excite to action."[84] The fictions of imaginative writers could be seen to promote immorality.[85]

What can we say of the fiction of natural philosophers? To begin with, we can say with certainty that fiction found a place in that empire of fact, natural philosophy, although natural philosophers generally avoided the word or used it disparagingly. Fiction in natural philosophy was similar to legal fiction, a falsehood contrived to fit existing law to recalcitrant reality. In the remainder of this section, we will consider several forms that fiction took in physical theory.

Scientific investigators used convenient idealizations, as the following example illustrates. In comparing his theory of gunnery with experiment, Benjamin Robins made an assumption that could not, strictly speaking, be true. It was that the action of the powder on a bullet ceases the instant the bullet leaves the barrel, whereas in reality the flame must continue to affect the bullet over some small distance from the end of the barrel. He supported his reasoning with the help of the following insightful statement on the practical limitations of the experimental confirmation of theories:

> It is well known, that in Experimental Subjects no such Preciseness is attainable; for those versed in Experiments perpetually find, that either the unavoidable Irregularities of their Materials, or the Variation of some unobserved Circumstance, occasion very discernable Differences in the Event of similar Trials. Thus the Experiments made use of for confirming the Laws of the Collision of Bodies have never been found absolutely to coincide either with the Theory, or with each other. The same is true of the Experiments on the Running and Spouting of Water and other Fluids, and of the Experiments made by Sir *Isaac Newton*, and for the Confirmation of his Theory of Resistances; in which, though they often differ from each other, and from that Theory by One-twentieth, One-tenth, and

> even sometimes One-fifth Part, yet those small Inequalities have never been urged as invalidating his Conclusions, since, in Experiments of that Nature, it was rather to be wondered at, that the Difference between the different Trials was so small.[86]

Without idealizations like that of Robins, it would be impossible to proceed one "Step in Natural Philosophy, since no Mechanic Problem hath ever been solved, in which every real Inequality of the moving Force hath been considered." Having compared the computed velocities of bullets with the results of a great many experiments made with guns of various sizes, Robins concluded, "By this Agreement between the Theory and the Experiments, each Part of the Theory is separately confirmed." Because experiments on "so furious" a force as that of fired gunpowder must vary one from the other, experimental deviations from the theory did not affect his conclusions.[87]

Consider another idealization, this from electricity. To facilitate the mathematical analysis of electrical systems, Cavendish connected electrified bodies theoretically by means of narrow canals of uniform matter filled with incompressible electric fluid, simulating conducting wires. Because in his theory the real electric fluid is elastic, the incompressible fluid in the canals "is not mentioned as a thing which can ever take place in nature, but is merely imaginary." Regarding this "ideal supposition" as the weakest part of his theory, Cavendish conducted experiments and made mathematical comparisons to justify it. It was later shown that his representation of "real canals" containing "real fluid" by imaginary "ideal canals" was mathematically as well as physically sound.[88]

From a general perspective, any mathematical discussion of a physical body could be viewed as an idealization. The instrument maker George Adams contrasted a "complicated and manifold" physical body with a mathematical body, which has only extension and figure, and thus is "purely *ideal*, the figment of mental abstraction."[89]

Models, a kind of idealization, allow parts of nature to be isolated, simplified, imitated, and manipulated; they ignore complications such as friction, which afterward are put back in. A theory might consist of a model together with a hypothesis about the similarity or analogy between the model and the world.[90] Newton showed that Kepler's laws of planetary motion are strictly true only in a model of the world in which a mass point moves toward a center of force; he then showed how Kepler's laws are modified to fit the real world.[91] After the mass point, other principal models of rational mechanics were a line with continuous mass, and a line loaded with discrete masses.[92] There were more complicated models; illustrating the propagation of force in bodies by a row of tiny balls connected by screw wires, Robison spoke of his "model," a rare appearance of the word in British writings on natural philosophy.[93]

There were practical as well as theoretical models, constructions such as a machine for illustrating the solar system, the orrery, and a machine for studying the power of water and wind. An inventor of the latter, the engineer John Smeaton, advocated the use of practical models in mechanical investigations but cautioned that it was always necessary to indicate how a model differs from the actual machine, for otherwise the "model is more apt to lead us from the truth than towards it."[94] The same would have been said of theoretical models in natural philosophy.

Closely related to models were thought experiments. Indispensable in theoretical physics today, they have been around a long time. Galileo made good use of them, as did, in the eighteenth century, Benjamin Thompson, better known as Count Rumford. Performing an "imaginary experiment," Rumford introduced a solid particle into a liquid mass and irradiated it with light, heating the particle and examining the result, proving his point that heat can be excited in a liquid body without immediately causing sensible effects. He explained his reasoning:

> The best method of proceeding in inquiries of this kind, where the principal object is to discover whether a supposed event, which, from its nature, cannot fall under the cognizance of our senses, is or is not possible, seems to me to begin by supposing the event to have actually taken place, and then to trace its necessary consequences, and compare them with those appearances which are actually found to have taken place.[95]

Trusting to established principles, Rumford's thought experiment was a theoretical analog of an instrument for extending the senses.

Confirmed theories of nature could be thought of as facts, and some philosophers have thought of them that way, or alternatively, they could be thought of as fictions. The twentieth-century theoretical physicist James Jeans said that contemporary physicists were concerned with "nothing more than pictures—fictions if you like, if by fiction you mean that science is not yet in contact with ultimate reality."[96] Natural philosophers tended to think of their work similarly; they did so if they followed Locke's example of a person imprisoned inside a diving bell, and who hears various signals but has no way of knowing which if any of the signals come from the outside. It is not obvious that, in this respect, eighteenth-century scientific realists were less sophisticated than their modern successors.

Laws of Nature, Principles

Natural philosophers placed high value on "laws of nature," another name for general effects constantly observed to take place under given circumstances.[97] There were a number of reasons for this. Natural laws compressed information, unburdening the memory. They were exact, preferably quantitative, inviting mathematical development. They joined experiment and theory: After deriving a law from experimental facts, the natural philosopher gave "a theory or explanation of the subordinate phenomena."[98] They predicted the future: Armed with "Laws of Electricity," the Royal Society's experimental demonstrator could "foretell what will happen to most Bodies, before the Experiments are tried upon them."[99] The starting point for reasoning about nature, they enabled the natural philosopher to "trace an established order, where a mere observer of facts would perceive nothing but irregularity."[100] Ubiquitous, timeless, they made sense of the seeming chaos of the world.

A working premise was that the laws of nature were "few and simple."[101] They were, according to some authors, the three "Laws of Motion," although that number was not absolute.[102] Other authors might exclude the third law of motion, as Cavendish intended to do in his treatise on mechanics, or they might add laws; one author, in a display of excess, added eighteen more "axioms" of motion.[103] The laws of motion were simple: Newton's second law, Cavendish said, was "the most simple & therefore the most likely to be true of any law one can invent."[104] The expression "laws of nature" was sometimes applied to general facts other than the laws of motion, but the latter remained its root meaning in Britain, one reason why in heat and other branches of natural philosophy synonyms for "laws" such as "rules" were found useful. Laws of nature were also sometimes referred to as "causes" or "powers," but that use was criticized on the grounds that "laws" do not stand for agents, but rather for ways in which agents act.[105]

Like laws, principles were a prized possession of natural philosophy. The early Newtonian James Jurin wrote, "Whatever is laid down on either Side [in a controversy] as a Principle, ought to be something all the World agrees in, at least what is admitted by the other Party."[106] In keeping with that broad characterization, "principles" normally referred to causes of activity in nature, and to starting points of reasoning in natural philosophy.[107] Newton, who used the word in the title *Principia*, did not clarify his meaning; in that book he apparently had in mind forces or laws of force.[108]

There were no hard and fast rules about the use of terms in scientific arguments, and natural philosophers used them interchangeably, a cause for complaint.[109] Depending on circumstances, a given statement could be, or be about, a principle, a theory, or a hypothesis. When called into question, a statement was likely to be called a hypothesis; when accepted, a principle or theory, though the categories could be combined, as in hypothetical principle. Drawing on his own experience, one investigator concluded that in any successful scientific inquiry, there comes a point at which "we balance the fertility of a principle, in explaining the phenomena of nature, against its improbability as an hypothesis."[110] Cavendish, a careful writer, in the same paper spoke of Lavoisier's antiphlogistic chemistry as a "principle," a "hypothesis," and a "theory,"[111] an ambiguity that reflected his effort at the time to understand just what Lavoisier's chemistry really was. An electrician who disagreed with Franklin's "hypothesis" of an electric fluid noted with dismay that it was "now almost universally received" in the philosophical world, "and, according to the opinion of some eminent electricians, it almost ceases to be a theory; and bids fair to be handed down to posterity, as equally expressive of the true principles of electricity, as the Newtonian philosophy is of the true system of nature." In this case, explanations of electricity were ranked in order of increasing authority: first hypotheses, then theories, and finally principles.[112]

Theories of Phenomena in the Laboratory

Consider two examples of theories drawn from laboratory physics: one a theory based on experimental principles, the other a theory based on a hypothesis. The first is a theory of friction proposed by the future Professor of Astronomy and Experimental Philosophy in Cambridge, Samuel Vince. Friction, the most neglected "branch of natural philosophy," was important alike for the "practical mechanic" and the "speculative philosopher," or theoretical natural philosopher. Previous theories of friction such as Leonhard Euler's "extremely elegant" theory had one basic problem: They were false, and for the reason that they were not founded on experiment. To replace them with a true theory, in part 1 of his paper Vince posed four queries about the principles of "friction as a force" and then answered them with experiments on the motion of bodies rolling down an inclined plane or projected on a horizontal one. In part Two, without recourse to a hypothesis, he established a mathematical "theory upon the principles" deduced from the experiments of part 1. The order of the paper followed the method of the experimental philosophy: Beginning with motions produced experimentally, Vince next stated the principles governing the force responsible for the motions, and then deduced further consequences of the force. His way of making a theory in connection with experiment was a textbook illustration of Newton's recommended method in natural philosophy.[113]

For our example of a theory built upon an hypothesis, we look at Cavendish's paper on electricity for 1771. Beginning the paper with "the method" he planned to follow, he stated the cause of electricity as a hypothesis: There exists an electric fluid the particles of which mutually repel but which mutually attract the particles of common matter, and the latter mutually repel; the distance dependency of these forces is the same, the inverse of some power less than that of the cube. Treating the electric force of a body as the sum of the forces of the elementary parts of the body, as Newton had treated the gravitational force, Cavendish built a mathematical structure for electricity in emulation of the *Principia*.

In part 1 of his paper, Cavendish laid down the theory of the electric fluid. In part 2, he compared the consequences of the theory with known experiments, noting agreements: This is a "necessary consequence of this theory"; "this is plainly conformable to theory"; these cases "appear to agree perfectly with the theory." Of his explanation of the Leyden jar, he said that he planned soon to be able to say "whether this agrees with experiment as well as theory."[114] Part 3, which Cavendish completed but did not publish, contained new experiments of his own, which were of two kinds: One refined the hypothesis, the other tested predictions of the theory according to the refined hypothesis. Deliberately incomplete in one detail, the hypothesis stated the law of electric force in such a way that it covered a range of possible distance dependencies and a corresponding range of consequences. Newton proceeded the same way in his investigation of the force of gravity, deducing different planetary orbits corresponding to different distance dependencies of the force. By the time Cavendish's paper was read to the Royal Society, he was reasonably confident of the correct law of electric force, the inverse square of the distance.[115] Later in the same year, he performed his so-called "hollow-globe experiment," which followed an unambiguous prediction of the theory: If the law of electric force is the inverse-square law, any redundant, movable electric fluid in a body lies entirely on its surface.[116] Confirming this prediction, Cavendish declared himself fully satisfied. In addition to deciding the law of electric force, the hollow-globe experiment confirmed the "truth" of the "theory" in general, for without the

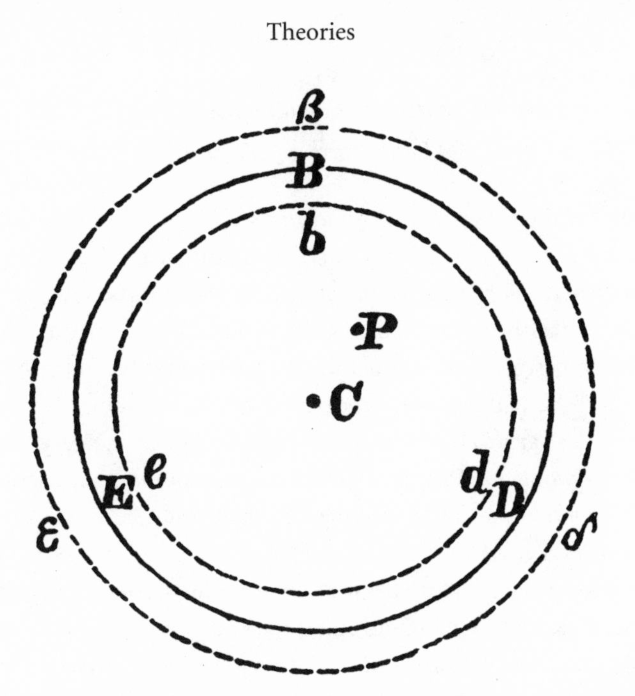

Illustration 1. Repulsive Force. The repulsion of a particle P by a spherical shell of repelling matter Bb depends on the the law of repulsion. If repulsion varies inversely as a higher power of the distance than the square, P is impelled toward the center of the sphere. If it varies inversely as a lesser power than the square, P is impelled away from the center. If it varies precisely as the inverse square, P experiences no net force. Citing a proposition in Newton's *Principia*, Cavendish developed this theorem on forces in his theory of electricity: "An Attempt to Explain Some of the Principal Phaenomena of Electricity, by Means of an Elastic Fluid," published in the *Philosophical Transactions* in 1771, and reprinted in *The Electrical Researches of the Honourable Henry Cavendish*, 8.

theory, the confinement of the redundant electric fluid to the surface of a conductor would not have been foreseen. Assuming the inverse-square law, he carried out experiments on electrical capacitances, providing further "great confirmation" of the "truth of the theory."[117]

Evidence for the truth of a theory increases with the precision of its explanations and the number of facts it explains, Playfair wrote.[118] By both measures, Cavendish's theory of electricity was successful. Through an experimental confirmation of phenomena predicted by the theory, he gave the hypothesis a mathematically precise formulation, and with the perfected theory he not only explained the previously known facts of electricity but also opened a new field of facts for quantitative experimentation: electrical capacitances.

Theories of Phenomena in the Laboratory of Nature

The methods of the experimental philosophy applied to subjects falling outside the experimental work of the laboratory. Using the method of analysis, with the help of mathematical principles, Newton deduced the law of gravity from the phenomena of the heavens, planetary motions. By the method of synthesis, using the same mathematical principles, he deduced other phenomena such as the tides, which he had not used in deriving the law, and in this way he further confirmed it.[119] That is how he and his followers saw his achievement. Without comment, his younger colleague Desaguliers incorporated gravitational astronomy into his course on the "Experimental Philosophy."[120]

As Newton had used the methods of analysis and synthesis in his study of the solar system, William Herschel used them in his study of the greater system, the universe. Herschel spoke of grouping stars according to their natural causes, applying the method of "analysing, if I may so express myself."[121] He opened his majestic paper of 1785 on the construction of the universe with a section, "Theoretical View," in which he discussed central forces, a consequence of which was the clustering of stars, the "*Laboratories* of the universe."[122] Four years later he recalled that in his "theoretical view," he had discussed star clusters in general. Should his readers have regarded it as "little better than hypothetical reasoning," he now discussed clusters in terms of observations, and then once again theoretically, as a consequence of the "action of central powers." In the "great laboratory of the Universe," his theory was secured.[123]

Other observational sciences made similar use of the methods of analysis and synthesis, availing themselves of the "great laboratory of nature."[124] Assuming that earthquakes originate in subterranean fires, John Michell explained that when pent-up water suddenly comes down on subterranean fires, steam is generated, the elastic force of which is greater than the force of gunpowder, more than sufficient to shake the earth. In presenting his theory of earthquakes, Michell followed Newton's example from astronomy by first describing the phenomena that led him to the "cause" of earthquakes, the "elastic force" of steam generated by subterranean fires; then by deducing the motions resulting from this force, waves propagated through the elastic, stratified substance of the earth; and finally by explaining other phenomena as effects of this motion.[125]

From a theoretical standpoint, observational sciences such as astronomy and geology were experimental sciences. Their practioners confirmed the effects of forces in a "laboratory," where truth in the experimental philosophy began and ended. The methods of the experimental philosophy established a unified basis for a gamut of physical sciences, in which sense they could all be included within natural philosophy, as they often were.

Predictions and Tests

Experiments and observations, if carried out in connection with a theory, were usually either to establish new facts as predicted by the theory or to test the truth of the theory, or possibly both. We encounter both purposes in examples from optics and gunnery.

As a preliminary, let us briefly review Newton's major optical finding, his proof that white light consists of colored rays of different refrangibilities. Determining that the latter differences were in large part responsible for imperfections in the images formed by refracting telescopes, and seeing no way to improve refractors other than by impractically lengthening them, he gave up on them and looked instead to reflecting telescopes.[126] The invention of the achromatic refracting telescope overcame his discouraging conclusion. Its inventor, John Dolland, found that a lens made of flint and crown glass could effectively eliminate chromatic aberration while retaining adequate refractive power. He believed that Newton had drawn an erroneous conclusion about dispersion from his theory of colors. Newton was defended by an author in the *Philosophical Transactions*, who chastised those who instead of assuming that Newton had made an error should have protected "so great a name," who instead of finding an absurdity in *Opticks* should have looked for a reading that was consistent with Newton's correct theory.[127] Another author conceded that there was an error in Newton's calculation, but not that Newton had made it. Rather, the text had to be "corrupt," and to make it right, he wrote a paper for the *Transactions* of the Royal Irish Academy.[128] Newton's "Theory," his own term, of light and colors had few if any serious doubters in Britain, although some specific theorems of his were found not to be universal.[129]

Newton was thought to hold a second theory, a theory about the cause of light. This theory raised a question at the outset: Was it even a theory? Priestley thought not. Newton did have a "theory" of light, his theory of light and colors, but his view of the cause of light was a "hypothesis," for it was "only proposed by him in one of his queries."[130] Not as fastidious as Priestley, his contemporaries tended to identify Newton's "theory" with what he said in the queries and to overlook the grammatical form in which he said it, a question: Light consists of "very small Bodies emitted from shining Substances."[131] Newton's readers were not far off the mark, for if he did not commit himself positively, he strongly leaned toward this explanation of optical phenomena. Widely thought to be true in eighteenth-century Britain, the particulate theory of light did have serious doubters, who looked to competing theories, one of which held that light is a fluid, another that it is vibrations of the aether, an interpretation that merged with the eventual mathematical theory of wave propagation.[132]

If, as Newton suggested, light consists of streaming bodies, it should behave like other bodies in all respects; in particular, it should possess momentum and respond to gravity and to powerful forces that come into play when light is reflected, refracted, or inflected by ordinary bodies. That understanding gave rise to a number of experimental investigations in the eighteenth century.

Our first example is the theoretical prediction of a new domain of facts. Combining Newton's theories of particles of light and of universal gravitation, John Michell devised an ingenious method for determining the distances, sizes, and weights of the fixed stars. On the assumption that light gravitates, and on the Newtonian hypothesis that the refraction of light is caused by a force impelling it toward a refractive medium, he reasoned from propositions in the *Principia* that light from a sufficiently massive star would suffer a detectable retardation by its attraction to the emitting star. Owing to its diminished velocity, the star's light would undergo a slight angular change when viewed through a prism or a lens. This change, a measure of the magnitude of the star, would allow the distance of the star to be estimated, solving the outstanding problem in the astronomy of the stars.[133] There was immediate interest, and Cavendish together with other astronomers looked for starlight exhibiting the diminished velocity. Failing to find it, they did not question the phenomenon, only the existence of stars massive enough for it to register with their best instruments. Received not as a test of the theory of light but as an opportunity for discovery, the prediction was of a kind that would characterize the coming age of precision. This unambiguous consequence of the theory was new, quantitative, and at, or just beyond, the limits of existing instrumental capability.

Let us now turn to tests of Newton's particulate theory of light. Theoretical principles might be confirmed directly, as happened with the principle underlying James Hutton's theory of the earth, an early instance of a laboratory investigation in an observational science,[134] but more often they were confirmed indirectly. That was necessarily the case with Newton's theory of light, because its particles were too small to observe.

The principles of the theory were tested by observing variations in the velocity of light. Newton's explanation of different refrangibilities by different sizes of

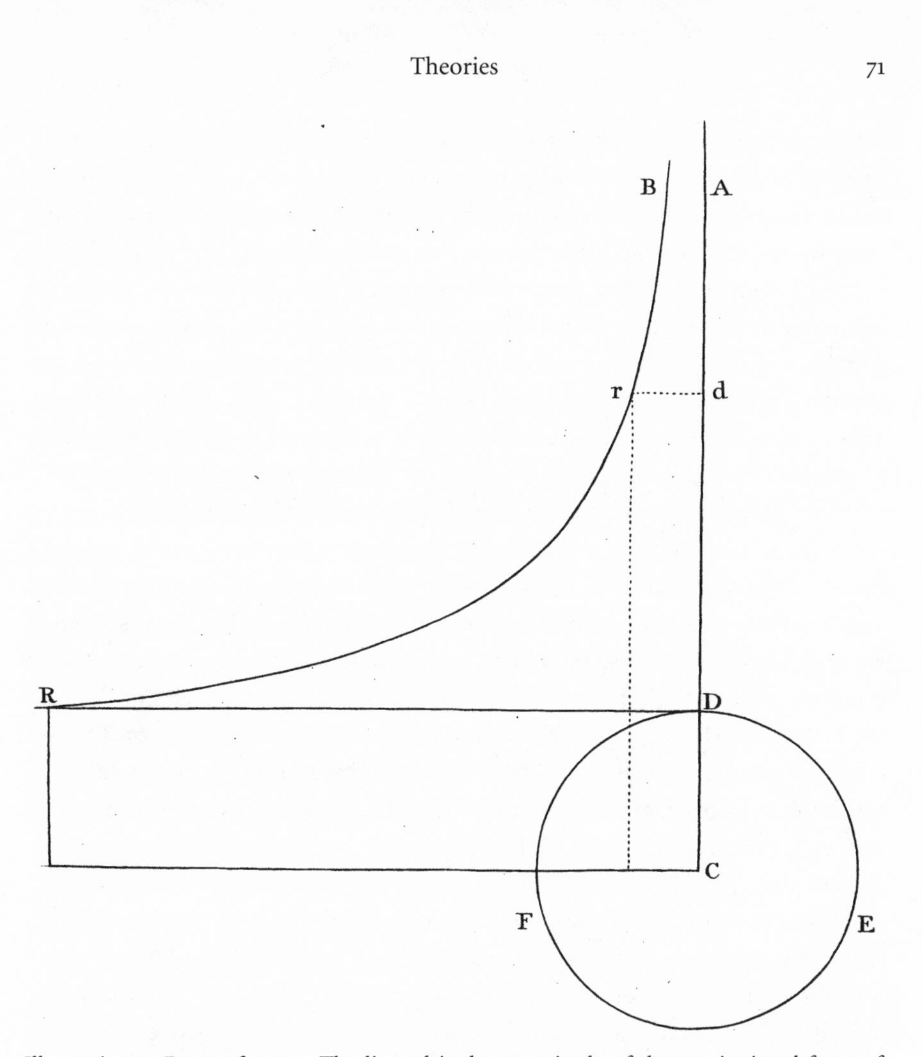

Illustration 2. Force of a star. The line rd is the magnitude of the gravitational force of the star DEF on a particle of light at a distance Cd from its center. This figure, which is rotated ninety degrees clockwise from the original, follows the last page of John Michell, "On the Means of Discovering the Distance, Magnitude, &c. of the Fixed Stars . . . ," *Philosophical Transactions* 74 (1784): 35–57.

the particles of light of different colors was disputed by Thomas Melvill, who accused Newton of making a false analogy between refracted light and falling bodies. If refraction is like gravitation, it should deflect light particles of every size exactly the same. Melvill proposed that the true cause of different refrangibilities is not different sizes but different velocities of the particles of light of different colors. The Royal Society ordered a test of this serious challenge to

Newton's theory. If red light travels fastest, a source of white light emerging from a shadow such as a satellite rounding the limb of Jupiter should appear red at first. Melvill's hypothesis did not stand up.[135] Thirty years later the matter did not seem closed to Vince, who believed that between the *Principia* and *Opticks*, Newton had indeed contradicted himself, that Newton's theory of light was inconsistent with his "Theory of Motion," and he urged Herschel to undertake observations of the occultations of stars by the moon to "settle the Theory of Light." It was important, Vince said, because "we so often apply the Theory of Motion to Light." Is light, he asked, "subject to the common Laws of Motion"?[136]

When particles of light pass through a refractive medium such as glass, they are accelerated, according to Newton, and the more refractive the medium, the greater is the acceleration. Patrick Wilson, assistant to the Professor of Practical Astronomy in Glasgow, wished to bring this statement to a test. His method was to observe the aberration of light, a minute periodic change in the apparent position of a star arising from the combined motions the earth and of the light from the star. The tube of a telescope was to be filled with liquids of different refractive powers; then if light moved faster or slower through the liquids, Wilson reasoned, the difference would be detected in the quantity of aberration, and the "Principles" underlying Newton's theory of light would be confirmed or denied.[137]

On the basis of speculations from Glasgow, Robison tried unsuccessfully to carry out an experiment with a telescope filled with water. He next designed a microscope to replace the telescope. He then examined the theory of a similar experiment proposed by Boscovich, who reasoned that a telescope filled with water and pointed at a terrestrial object would not stay pointed at the object but would deviate at a definite rate. This terrestrial version of aberration had striking implications: With a telescope of Boscovich's design, an observer in a dungeon could determine the motions of the earth and the sun or make astronomical discoveries. A proper experiment on this subject, Robison believed, could decide "that important question in physics," which of the two principal hypotheses about light was correct, Newton's particles or Euler's vibrations. If the telescope always pointed at the object regardless of the liquid it contained, the answer would favor Newton's hypothesis and the experiment would amount to "almost a demonstration that light consists of corpuscles emitted by the shining body." Robison expected the latter outcome from his own theory of the experiment, which differed from Boscovich's. Nothing conclusive came of this line of research.[138]

For our last example from optics, let us return to Michell. To demonstrate that light has momentum, and to measure its quantity, he concentrated sunlight on a copper vane suspended from a counterpoised wire. Observing a rotation of the vane in the direction of the motion of light, he concluded that the vane moved in reaction to the momentum of the light striking it. Just how satisfied he was with the experiment is unclear, but it was widely received as proof that light is swiftly moving particles of matter, the principle of the Newtonian theory.

Priestley used Michell's experiment to refute a well-known objection to Newton's theory. If light consists of an outpouring of innumerable particles of matter from the sun, as Newton maintained, then the sun must waste away, with the result that its gravitational pull on the earth must diminish, altering the earth's motion. Michell's experiment implied that in its six thousand years, the sun had lost only 670 pounds of its matter in the form of light, too little to be observed. By another line of reasoning, Samuel Horsley calculated that in many hundreds of millions of years, the sun would have lost less than a hundred thousandth part of its matter, again too little to observe.[139] Michell, Priestley, and Horsley evidently showed that Newton's theory of light did not stand in contradiction to his theories of motion and gravitation. For the time being, for most British authors, light remained swiftly moving, extremely small particles of matter, and the Newtonian philosophy remained unshaken.

In due time, as all good theories, Newton's theory of light was replaced. Michell's experiment on the momentum of light, originally a support of the theory, was turned against it. Abraham Bennet repeated the experiment with a more sensitive apparatus, a small round piece of paper fixed to a counterpoised horizontal arm suspended from a spider's thread inside a glass cylinder. Upon exhausting the cylinder and directing a ray of light at the paper, Bennet observed no movement of the arm. What Michell had seen and explained by the momentum of light, Bennet explained by the greater heat of the air on the illuminated side of Michell's vane, the expanding air pushing the vane in the direction of the ray of light. Having designed his experiment to test a consequence of Newton's theory, the mechanical impulse of light, and finding no evidence of it, Bennet offered his experiment as support for an alternative theory of light: "[H]eat and light may not be caused by the influx or rectilineal projections of fine particles: but by the vibrations made in the universally diffused *caloric* or matter of heat, or fluid of light."[140]

Thomas Young considered Bennet's experimental disproof of the Newtonian theory convincing and his theoretical conclusion correct. On the basis of an analogy between light and sound, Young argued for a wave theory of light similar to Huygens's and Euler's. At the same time, Young acknowledged a debt to Newton, who had suggested a wavelike property of light, and whose experiments on thin plates had helped persuade Young of the wave theory. In natural philosophy, Young said, Newton was "great beyond all contest and comparison," but his doctrine of the particles of light was wrong. To develop what Young called "my theory," he laid down hypotheses about an elastic, luminiferous aether pervading the universe.[141] Alternatively referring to the wave theory as a "hypothesis," he left it to other physicists to secure its status as a theory in the nineteenth century.[142]

Our final example of experimental testing of a theory is taken from gunnery, an active field of research at the time. The mathematician and military engineer Benjamin Robins was, according to an artillery officer of the time, "in gunnery what the immortal Newton was in philosophy, the founder of a new system deduced from experiment and nature."[143] The president of the Royal Society agreed, attributing to Robins a "new science" of gunnery[144] and in 1746 awarding him the Royal Society's Copley Medal for experiments that "confirm and ascertain his Theories."[145]

According to the old theory of gunnery, projectiles follow a parabolic path, but practical experience and scientific experiments taught otherwise. Small projectiles fired at high velocity through the air were found to travel only one tenth or one twentieth as far as they would in empty space. The path of a projectile is further complicated by an oblique action of the air on it, which was thought to arise from a rotation caused by the friction of the bore on the projectile.[146] When the great force of air resistance is taken into consideration, the actual motion of projectiles "becomes one of the most complex and difficult problems in nature."[147]

Robins attributed the force of gunpowder to an elastic fluid condensed in the powder, which upon ignition expands violently and propels the bullet; he measured this fluid and force experimentally. From his theory of gunpowder together with a proposition from the *Principia* on the motion of bodies under central forces, he related the force of the gunpowder acting on the bullet along the length of the barrel to the square of the velocity of the bullet.[148] From this result, he readily deduced a law expressing the velocity of a bullet in terms of several measurable factors: the weight and specific gravity of the bullet, the dimensions of the bore, and the length of the charge.

To test his law, Robins invented a simple, accurate instrument, a ballistic pendulum using a large block of wood as weight and target. In what an admirer called a "beautiful military experiment," Robins fired bullets into the block and observed its reaction. Knowing the weight of the block, the distances of its center of gravity and center of oscillation from the point of suspension, the weight of the bullet, the point where it struck, and the arc through which the pendulum swung, he deduced the velocity of the bullet in what was for the time a sophisticated mechanical analysis. Finding a close match between the experimental velocity and the velocity calculated a priori from his theory, he confirmed the principles of the theory. By firing at the pendulum from different distances, he deduced the resistance of the air, completing the theory of gunnery to his satisfaction.[149]

Both "theoretically and practically," Robins showed that the resistance to bullets is "enormously great, much beyond what any former Theories had assigned."[150] The investigation was carried further by the mathematics professor at the Royal Military Academy at Woolwich, Charles Hutton, who agreed that Robins had proven his theory as far as it went, but that to form a "proper theory," experiments using bigger shot than musket balls had to be performed. Having at his disposal ranges for firing cannon, and with the help of Robins's pendulum for high velocities and Robins's whirling machine for low velocities, Hutton measured the resistance of air over a wide range of velocities, from 0 to 2,000 feet per second, from which he concluded that the resistance varies according to the 2 ¹⁄₁₀th power of the velocity, close to Newton's theoretical law. He found that the force of gunpowder has twice the strength of Robins's estimate, but otherwise his results confirmed Robins's general theory. Hutton observed that knowledge of the velocities of balls shot from cannon and of gunnery in general was useful not only to gunnery, a practical application of rational mechanics, but equally to theoretical natural philosophy.[151]

Rumford, a man of science and invention with a lengthy military career, tested Robins's theory by measuring velocities using the pendulum method, and also using the method of the recoil of the gun. He concluded that the "principles upon which [Robins's] theory is founded are erroneous," in the process establishing experimentally the true law of velocity. Robins's champion Hutton did not accept Rumford's law.[152] Later, expanding his criticism of Robins's theory, Rumford measured the force of gunpowder directly, finding it fifty times greater than Robins's figure. With such a discrepancy, he asked, "[w]hat will become of this [Robins's] theory, and all of the suppositions upon which it is founded"? Moreover, "all the theories hitherto proposed for the elucidation of the subject, must be essentially erroneous."[153] Gunnery offers an instance of extensive interplay of theory and experimental testing in the eighteenth century.

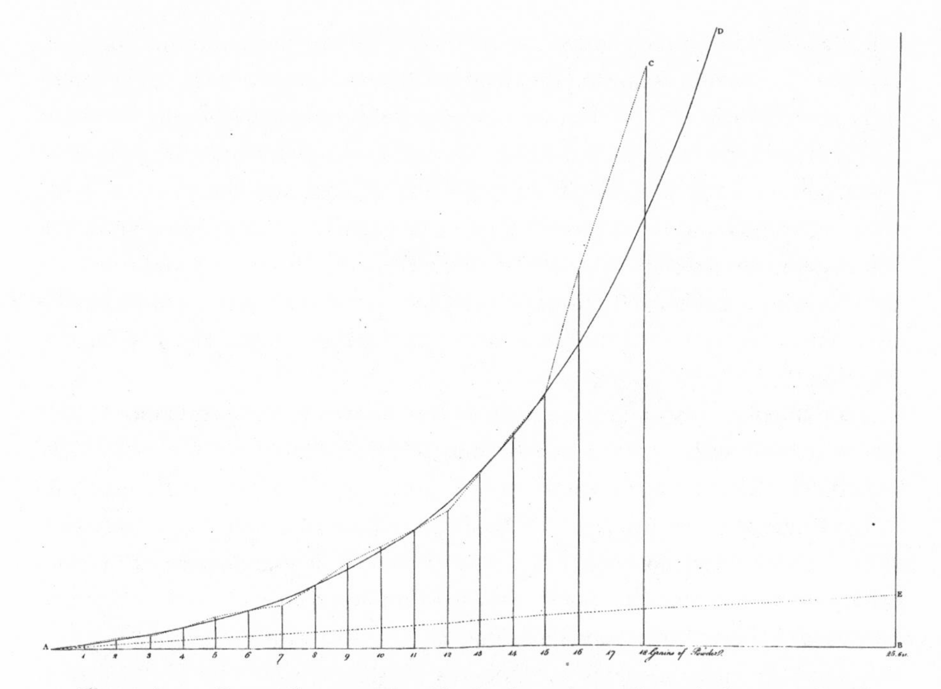

Illustration 3. Force of gunpowder. The horizontal coordinate is the quantity of gunpowder or, the same thing, the density of the elastic fluid generated by the fired powder. The vertical coordinate is the elastic force of the powder. AC is the experimental curve, AD the mathematical law. The straight line AE is the law assumed in the standard but incorrect theory. Embedded in the tabulated data, the correct law "may be seen in a much more striking manner by a bare inspection of the figure." Benjamin Thompson, "Experiments to Determine the Force of Fired Gunpowder," *Philosophical Transactions* 87 (1797): 222–92.

Desire for Theory

In Cavendish's time, in one after another scientific field there were calls for theoretical understanding, as is clear from the following statements. Because of the imperfect development of instruments for measuring the atmosphere, "all theoretic inquiries" had long been impeded.[154] A mid-century observer of the weather looked to the Royal Society and its journal to facilitate "more perfect Theories of Wind and Weather" or, because of the capriciousness of the weather, to expose the vanity of all such effort.[155] At the end of the century, a call was made for registers of winds and weather around the globe to establish a "proper theory."[156] The wait would be long. Nearly as daunting as the weather, the land

and the sea would be understood only when the earth acquired a truly scientific "theory."[157] Priestley thought that the principles of Franklin's electrical theory were as sound as those of Newton's theory of gravitation, but he also thought that the experimental base was too narrow, and he called for more and varied experiments to achieve a "perfect general theory."[158] The author of an electrical explanation differing from Franklin's did not propose a theory, but he trusted that his alternative principle would further the goal of a "complete and consistent theory of electricity."[159] A writer on the variation of the compass needle advocated making numerous observations at many places on the globe as the only way to "arrive at a true theory."[160] The inventor of a new instrument to measure the dip of the compass needle proposed its use to determine the earth's magnetic poles and, in that way, to contribute to the goal of earth magnetism, to "complete the magnetic theory of this globe."[161] Bennet anticipated that researchers using instruments as delicate as his would one day reach the ultimate goal of the study of magnetism, its "true theory."[162] The *Principia* gave an entirely new face to "theoretical astronomy," the Cambridge astronomer Roger Long said.[163] To a mathematically adept natural philosopher such as Cavendish, the object now was to give the other theoretical parts of natural philosophy *Principia*-like faces. Cavendish's desire for a true theory of heat and the way he went about it fit a pattern of expectation.

A Great Question

Historical Setting of Heat and Mechanics

Heat

The history of heat is itself a chapter in the history of natural philosophy and cannot be fully understood apart from it. Our detour through natural philosophy in part 1 has prepared us for the subject of part 2.[1]

Let us begin with a brief overview. Heat, as it was understood in the eighteenth century, belongs either to the sense of touch or to a distinct sense. In either case, the sensation of heat is relative: Loss of heat from our body to another body occasions our sensation of cold; gain of heat occasions warmth. We perceive no heat in bodies if they are at the same temperature as our own, although we still speak of the thermometer as reading "sensible heat." Our sensations, "imperfect and deceitful measures of heat,"[2] are too imprecise to rely on in science, and in their place we substitute the effects of heat, which can be reduced to the expansion of bodies with heat, the principle of the thermometer.

As much as in any field of science, the understanding of heat developed in step with the development of an instrument. Invented in the seventeenth century, the thermometer evolved into an instrument of considerable precision over the period covered in this book. By the middle of the eighteenth century, the calibration of thermometers between the freezing and boiling points of water had become commonplace. When a committee headed by Cavendish found that the boiling point of thermometers owned by the Royal Society varied by two or three degrees Fahrenheit, it issued firm recommendations to ensure uniformity in marking this calibration point in the future. A correction for the dilation of glass made further progress toward comparability in thermometers. By the 1770s, thermometers could be read to one or two tenths of a degree, and before the century was over, at a stretch to one one-hundredth.[3] Mathematically expressed, Cavendish's understanding of heat was well suited to the quantitative stage of contemporary researches in heat. It is fitting that this expert on the thermometer, who made good use of it in his researches on specific and latent heats, should be the first natural philosopher to work out a complete mechanical theory of heat.

Newton on Heat

"Newton's theory," Cavendish wrote in the manuscript "Heat." Choosing his words carefully, he first wrote Newton's *hypothesis*, then crossed it out and wrote Newton's *theory*. We begin by looking at what he meant by Newton's theory of heat.

Whether or not Newton thought he had a theory of heat, he had a good deal to say on the subject. To measure heat, he proposed a scale, with ice at 0, the human body at 12, and burning coals at 192. For low temperatures, he used a linseed oil thermometer; for high temperatures, he used a plate of red-hot iron, counting the time from any given instant until the plate cooled to the temperature of the human body. The latter method depended on a law of cooling, his own law, which stated that the rate at which bodies cool is proportional to the difference between their temperature and that of their surroundings.[4] In the *Principia*, within a discussion of comets, Newton reported that a globe of red-hot iron an inch in diameter cools in about an hour, and on the basis of geometry, he estimated that a bigger globe takes longer, in proportion to the ratio of the diameters; he suspected that other causes were active too, and he suggested that experiments be made to decide this question, as they were.[5] A notable contribution to thermometry, Newton's law of cooling was useful to Cavendish.

Newton's most extended discussion of heat occurs in *Opticks*. Queries 5–11 are about the intimate connection of heat and light. Here, as part of a question, we read Newton on the cause of heat: Light acts upon bodies by "putting their parts into a vibrating motion wherein heat consists."[6] The rays of light and the parts of ordinary bodies interact by forces: When light is absorbed by a body, by the law of action and reaction, its parts are caused to vibrate, registering as heat; when the vibration is sufficiently violent, particles of absorbed light are removed from the reach of attraction and enter a region of repulsion, there to be "shaken off" and driven from the body with "exceeding great Velocity."[7] Similarly, Query 29 says that the interaction between rays of light and bodies "very much resembles an Attractive force between Bodies," causing the parts of bodies to vibrate, heating them; and Query 31, the final query, gives abundant examples of heat accompanying chemical changes in bodies, all of which are explained by the vibration of particles.[8] In between, Query 18 conjectures that bodies communicate heat by vibrations of a subtle, elastic medium, the aether.[9] Newton made further observations on heat in the text of *Opticks*, scholia of the *Principia*, the *Philosophical Transactions*, and elsewhere.

In their totality, Newton's many scattered references made a forceful case for the significance of heat in natural philosophy. Different bodies have different heating properties, he noted. Black bodies are heated more than bodies of other colors. Large bodies preserve heat longer than small ones, the sun and the fixed stars longest. Heat liquifies solid bodies and vaporizes liquids. Mixtures of water and oil of vitriol (concentrated sulfuric acid), and of iron filings and aqua fortis (nitric acid), generate heat. With vivid illustration, Newton showed how interacting particles produce large, at times devastating, effects, attended by heat. When attracted to one another, particles "clash with great violence, and grow hot with the motion, and dash one another into pieces, and vanish into Air, and Vapour, and Flame." By the same cause, particles undergo "great and violent" motions, producing tempests, hurricanes, landslides, boiling seas, thunder and lightning, and fiery meteors. And by the same cause, "Bodies burn and shine, Mountains take fire . . . and the Sun continues violently hot and lucid, and warms all things by his Light." In the bowels of the earth, particles of sulfur rush toward particles of minerals, and "if pent up in subterraneous Caverns, burst the Caverns with a great shaking of the Earth."[10] Heat causes these phenomena and much more, just as, conversely, heat is caused by all kinds of natural processes: friction, percussion, putrefaction, vital motion, fermentation, combustion, chemical action, electricity, and actions in the interior of the earth and in the sun. Little wonder that eighteenth-century authors, admiring of Newton and well read in the queries of *Opticks*, were encouraged in their belief that heat is a driving force of the physical world, in need of experimental and theoretical elucidation.

Specific and Latent Heats

Because of the importance of specific and latent heats for research on heat in the second half of the eighteenth century, I devote a separate discussion to the subject. If this seems labored, it is nevertheless necessary for our understanding of Cavendish's work on heat.

During Cavendish's time, the pioneering researches in heat were done by William Cullen and Joseph Black. The older of the two, Cullen was Professor of Medicine and Lecturer in Chemistry at the University of Glasgow, in whose laboratory Black worked for a time, and whom Black succeeded at Glasgow as chemical lecturer. Cullen moved to the University of Edinburgh as Professor of Chemistry in 1756, in which position Black again succeeded him ten years later. Cavendish knew neither of them, but he was well aware of their work.

Historians observe that Cullen's and Black's general outlook on science was shared by Scottish Common Sense philosophy, the main assumptions of which were that there exist first principles, or natural laws, which are incapable of a priori demonstration, and that all knowledge arises by reasoning from these principles. The Common Sense philosopher Thomas Reid denied that hypotheses have any place in natural philosophy, and he limited the role of theories to deductions from experiments and observations. His colleague Dugald Stewart admitted a limited role for hypotheses. Wary of hypotheses, drawn to Newton's method more than to his corpuscular ideas, Scottish natural philosophers saw their task as the establishment of principles or general laws, although they also took an interest in physical causes.[11] As practiced by Cullen and Black, Scottish natural philosophy was philosophically sound, scientifically prudent, and perhaps overly cautious.

Stimulated by the simple observation of a student that a thermometer cools when it is removed from a solution, and recalling a similar observation by a French physicist, Cullen suspected that evaporation was the cause, and he conducted experiments to find out. Evaporating a number of solutions, he produced cold of "so great a degree" that he thought it could not have been observed before. For this reason, he urged that the whole subject be "further examined by experiments." He published his own experiments in 1756.[12]

Perhaps as early as 1757–58, Black lectured on the heat accompanying a change of state of matter. To convey his understanding, he gave a persuasive counterexample: If snow and ice were to melt immediately upon adding a small quantity of heat at the melting temperature, as was "universally" believed to occur, every spring the world would suddenly be overwhelmed by dreadful floods, which "would tear up and sweep away every thing, and that so suddenly, that mankind should have great difficulty to escape from their ravages." The reason why this devastation does not occur is that it takes time for ice and snow to absorb the heat originally lost in the change of state of water to ice and snow. In 1760, Black began to experiment on his own, and 1761 he measured the heat of fusion of ice, which he reported to the local scientific club in Glasgow the following year. Later he extended the concept to heats involved in changes of state of other substances, and he mixed ice with various salts, melting the ice and producing cold.[13] He coined the term "latent heat," standing for the heat absorbed or produced in a change of state.

At Black's request, in the early 1760s his assistant William Irvine determined the heat absorbed in melting metals such as tin, and the heat absorbed in melting soft substances such as spermaceti and beeswax. In 1764, with Irvine, Black obtained a value for the latent heat of steam upon condensing water vapor in a worm tube in a cold-water bath.[14] This experiment Cavendish learned of and repeated, and he may have learned of the experiments on tin and other metals, and also on spermaceti and beeswax, as he carried out experiments on these same substances as well.

The earlier opinion that bodies hold and exchange heat in proportion to their mass, volume, or density was wrong. Experiments from the first half of the century were reasonably convincing on this point. Different kinds of matter communicate heat differently, "for which no general principle or reason" had been given, Black said. His own reason, which came to him after reading George Martine's essay on rates of heating and cooling and Herman Boerhaave's book on the elements of chemistry, which reported relevant experiments by Daniel Gabriel Fahrenheit, was the concept of specific heat, or "heat capacity," the measure of the heat required to raise the temperature of a given weight of a specific substance by one degree. This was a new, permanent, and characteristic property of substances. He came upon this second, centrally important concept of heat in 1760.[15]

Historians of science attribute to Black the quantitative science of heat, known as "calorimetry." The first to appreciate the distinction between temperature and quantity of heat, he showed how to measure the latter by the former. Historians also regard him as the "first investigator to formulate theoretical explanations of these phenomena and to test his theories by the necessary quantitative experiments."[16] Black himself was cautious about speaking of a theory, but his contemporaries referred regularly to his theory, or doctrine, of latent heat.

Irvine followed Black's direction, but not his explanation. His experiments on spermaceti and beeswax suggested to him that their gradual change of state was best explained by a gradual change of heat capacity accompanying a change of form. Latent heats, in his view, arise from differences in the heat capacities of the solid and liquid states of a substance; the absorption of heat is the consequence of melting, not the cause. Ice absorbs heat in melting because at the melting temperature, the heat capacity abruptly increases. Black thought that Irvine's theory was not an explanation, because it did not explain the change of form.[17] At the level of the phenomena, a theory of heat was an incomplete theory for both Irvine and Black, as it was for Cavendish, who looked for a physical explanation.

Irvine's "theory" was characterized by a few simple ideas. Each body contains a certain quantity of "absolute heat," as well as a certain quantity of "sensible heat," as registered by the thermometer. Bodies at the same temperature contain different quantities of absolute heat depending on their nature and weight. Specific heats are proportional to absolute heats. Around 1770, Irvine developed a method for determining absolute heats from specific and latent heats, from which he derived an absolute zero, around minus 900 degrees Fahrenheit, the same for all bodies.[18] His ideas were sharply criticized by A. L. Lavoisier and P. S. Laplace, who on the basis of measured specific heats found that different substances give greatly divergent absolute zeros, and that quantities of heat in bodies are not proportional to their specific heats. Despite this and other evidence against it, together with the little evidence for it, Irvine's theory enjoyed a considerable following into the nineteenth century. A major reason for its long success was the absence of new theoretical approaches and the continuing unreliability of much of the data of heat.[19]

The scientific world learned about the new theories of heat indirectly, since neither Black nor Irvine published anything on them. Black's understanding of heat first became known through his lectures, student notes of which were in circulation by 1767.[20] An anonymous account of his lectures was published in 1770, and two years later a student published his ideas on heat in a French scientific journal. Irvine's papers were published by his son after his death, but long before that his opinions had become public. By the late 1770s, any serious investigator of heat, Cavendish among them, would have known about Black's and Irvine's researches in some detail.

Black's refusal to publish annoyed and confused some of his colleagues, who were trying to clarify the subject and the question of priority, but he also had his defenders.[21] His students Adair Crawford and William Cleghorn were among the first to make his ideas widely accessible, around 1780. P. D. Leslie and J. H. de Magellan also published on Black's work about this time, and the second edition of the *Encyclopaedia Britannica* came out with an account. Crawford expressed his hope that Black and Irvine would yet publish their discoveries,[22] although at this late date, they would mainly have confirmed what the scientific world already knew from other sources.

Cause of Heat

Heat had three main explanations, according to Isaac Milner, Jacksonian Professor of Natural Philosophy in Cambridge. We know that one of them was the vibrations of the parts of ordinary bodies. In this explanation, fire is any body heated sufficiently to give off copious light. Fire as a special kind of body was the starting point of the other two explanations. In one version, when fire is in a free state, its particles move rapidly in all directions, their motion registering as sensible heat; the particles lose their motion when they combine with particles of ordinary bodies, in which state fire loses its property of sensible heat. In the other version, the particles of fire do not enter into combination with those of ordinary bodies, and fire always retains the property of sensible heat; because different bodies have different capacities for heat, an addition of the same quantity of fire results in different increases in sensible heat.[23] The three explanations were often referred to as two, the mechanical, motion, or vibration theory, and the material or fluid theory. If fire were identified with the aether, the two explanations could be seen almost to coincide,[24] but normally they were believed to be in competition.

Black thought that heat is probably a special matter, which forms a union with ordinary matter, much as an acid does with an alkali. By this interpretation, he gave a new explanation of what Cullen had observed, the cold produced by evaporation: The heat absorbed by water in converting to vapor is not lost but is retained in the vapor, with which it combines, and to which it imparts elasticity. In agreement with the prevailing Newtonian world view, Black explained the union as the bonding of particles of heat with particles of vapor by chemical forces of attraction.[25]

On the question of heat, Irvine was undecided. He rejected the term "latent heat," since it suggested a special matter, preferring to speak noncommittally of the "cause of heat." "Heat capacity" also suggested a substance, and although at first Irvine used the term, he replaced it with "relative heat." Soon there appeared other neutral terms, "comparative heat" and, still in use, "specific heat."[26]

The view of heat as a special kind of matter was the subject of William Cleghorn's inaugural dissertation at Edinburgh University in 1779. He assumed that there exists a subtle and indestructible fluid, a fluid sui generis, the matter of fire, which is activated by "tried and tested" Newtonian principles, repulsive and attractive forces. The particles of fire mutually repel, imparting elasticity to the fluid, and are attracted to the particles of ordinary bodies, the strength of the attraction varying from substance to substance. The "equilibrium" of heat between bodies is brought about by these forces. Laying down "principles" of fire, the cause, Cleghorn deduced the "effects," eight in number: fluidity and evaporation, inflammability, animal heat, heat from electricity, fermentation, friction, mixtures, and the sun's rays. He gave a few simple formulas for comparing the absolute heats of two bodies, but because he did not know the laws of the forces he introduced, he could not develop his theory mathematically. In keeping with the Newtonian tradition, he concluded with a passage from Newton's final query in *Opticks* on particles and forces, followed by queries of his own on the same topic. What, he asked, are the "laws" of the repulsive and attractive forces of the particles of fire?[27] Black thought that Cleghorn's explanation of heat was the best to date, although he had reservations about it similar to the ones he had about Irvine's.

The question, "so much agitated among philosophers," of the nature of heat led to attempts to weigh the fluid matter of heat and in that way establish its existence. The Scottish chemist George Fordyce announced that upon melting, ice loses weight, and since when ice melts, it absorbs heat, heat possesses not weight but levity. This finding was not unreasonable, since heat was known to diminish electrical and other attractions. Fordyce's subsequent experiments persuaded him that heat is not a levitating substance after all, nor any kind of substance.[28] Initially a believer in heat as a substance, Benjamin Thompson, Count Rumford was intrigued by Fordyce's experiments. Upon repeating them in 1787, he decided that bodies do not change weight when they change state, and he came to doubt that heat is a substance.[29] Black told his students that the failure to prove that heat has weight was a "strong objection" to the view of heat as a substance, but he thought that the effects of heat—fluidity, expansion, vaporization, incandescence, and combustion—overcame the objection.[30]

The reasons given for adopting the fluid theory of heat were not compelling to everyone, but the theory had much to recommend it. Having developed together with experiment, it agreed with the facts;[31] it was in step with fluid theories in other departments of natural philosophy;[32] and it was readily grasped, plausible, predictive, and supported by leading authorities of the day.

The opposing view, the theory of heat as the internal motion of bodies, had wide currency in the seventeenth century, supported by Bacon, Boyle, and Newton. Black no doubt had this formidable company in mind when he said that contrary to the "greater number of the English philosophers," he thought that heat could not be motion, that motion is "totally inconsistent with the phenomena" of heat.[33] To Black, as to many others, it seemed that mechanical ideas of heat had not kept pace with the experimental development of the science in the late eighteenth century.[34] Their criticisms would have discouraged all but the most determined of their contemporaries from trying to formulate a mechanical explanation of heat.

In lectures in the mid 1780s, Issac Milner marshalled all of the main objections to the theory of heat as the internal motion of bodies. The motion of particles is unproven, he said. Heat passes slowly through bodies, rather than rapidly, as does motion. Heat diffuses, although it should not according to the motion theory, because the momentum of a system of particles is unaffected by their mutual actions and collisions. Heat is transmitted across a vacuum, although there are no intervening particles to set in motion. The liberation of heat during freezing is unintelligible as motion, as is the generation of cold upon evaporation. Milner, as it happened, was not overly impressed by these and other objections. A die-hard supporter of the motion theory, he thought that although the "arguments against this Theory have of late Years been esteemed so numerous and weighty that it has almost been given up by Philosophers," it had been given up "a little too precipitously." Conceding that heat is not observed to be proportional to motion, he restricted heat to one kind of motion, vibration, moreover, to vibration of a "particular kind," not further specified. With that understanding, Milner proceeded to make a case for the motion theory of heat.[35]

Rational Mechanics

At the same time that Milner was telling his students that no one "had endeavoured to shew the truth" of the vibrational theory of heat by contrasting it with the fashionable material theories, Cavendish was working on just such a proof. For this, he needed "rational mechanics."

Rational mechanics was the "science of motions resulting from any forces whatsoever," according to Newton.[36] It was also called "dynamics," Leibniz's term, as distinguished from "statics," or practical mechanics, concerned with mechanical powers such as the lever.[37] In principle, rational mechanics held dominion over all parts of natural philosophy, imparting to it a conceptual unity. Because "nearly all of the phaenomena of nature are owing to motion . . . the laws of motion must be looked upon as the foundation of natural philosophy," a text on natural philosophy stated.[38] Another stated that all physical phenomena arise from "ultimately mechanical forces, producing local motion and changes of motion."[39] Texts on natural philosophy began accordingly with objects of interest in mechanics: forces, bodies, and the laws of motion and of arrested motion.[40]

The authority of mechanics in natural philosophy derived from its success, scope, reasoning, and evidence, and perhaps also from the empiricist theory of knowledge. Its success was most evident in the mechanics of the solar system, the system of the world. In scope, mechanics was unlimited: If another force were to be discovered, its effects would be known through the laws of motion; on the day that the force responsible for, say, combustion became known, a "new field of mechanics" would be opened up.[41] Based on exact laws, the mathematical theory of mechanics was the model of accurate thinking in natural philosophy. Usually treated as empirical principles or axioms, the laws of motion were deduced from evidence by the "strictest philosophical reasoning,"[42] although they were occasionally deduced from definitions of space and motion or philosophical axioms,[43] a foreshadowing of our theoretical physics, which derives laws of nature from more fundamental principles. The evidence for the laws of mechanics was of three kinds: first, the everyday experience of humanity; second, experiments in which "every power is removed as far as may be, except those which are the objects of examination"; third, a posteriori arguments in which the theory of mechanics is deduced from the laws of motion and is found to agree with the facts without exception. Taken altogether, the evidence for the truth of mechanics was "scarcely inferior" to mathematical demonstration. The laws of motion had consequences of "such coherence and consistency among themselves and with matter of fact, as are rarely to be found in other branches, which admit not so intimate an union with the science of quantity." If every imperfection of an experiment were removed, and if the senses were infinitely improved, "mathematical coincidence" would obtain between the laws of mechanics and experiment. Mechanics was a "mathematical branch of physics."[44]

Vis Viva

To place the theory of heat on mechanical foundations, Cavendish did something unexpected of a British natural philosopher: He proceeded from a doctrine originating with Newton's German countemporary and arch-rival, the coinventer of the calculus, G. W. Leibniz. The doctrine in question was that of vis viva.

Leibniz made a distinction between "living force" and "dead force," vis viva and vis mortua. Dead force strives to generate motion, the measure of which is pressure, as in Newtonian mechanics; it is also potential vis viva. Commonly called the "force of moving bodies," living force resides in a moving body, generating continuing motion, overcoming obstacles, and communicating motion in collisions. Measured by the product of the mass and the square of the velocity, vis viva is an undirected, absolute quantity. It obeys a conservation rule, its most important property.

Unless they are students of the history of science, today's readers are unfamiliar with vis viva, although probably not with our related concept of kinetic energy, one-half the product of the mass and the square of the velocity. The missing factor one-half goes back to Leibniz, who introduced vis viva in a criticism of Descartes's measure of force, quantity of matter multiplied into velocity. In practice, his reformulated measure normally required multiplying by one-half.[45]

Keep in mind that the counterpart of vis viva in Newtonian mechanics is a directed quantity, the product of mass and velocity, or momentum, and that it, too, obeys a conservation rule: In the collision of any two bodies, the sum of their momenta in the same direction remains constant. Newton's hand-picked experimenter in the Royal Society, Desaguliers, included this rule under the "Laws of Nature," designating it "the universal *Law of Motion*."[46]

Originally thought to express God's design in preserving His Creation,[47] vis viva was assumed incapable of being destroyed without giving rise to a comparable effect. This understanding was well suited for the treatment of a range of mechanical problems, but not of collisions between bodies; from experience it was known that collisions are never perfectly elastic, from which it follows that motion and vis viva are always lost. Because for Leibniz and his followers, belief in the conservation of vis viva was firm, the missing vis viva was regarded as only apparently lost, continuing on as a hidden effect such as the compression of bodies or the motion of parts internal to bodies. Leibniz proposed the latter explanation, but he did not identify the hidden vis viva with heat, even though in his day heat was commonly believed to be the internal motion of bodies. It would seem that the conceptual problems of regarding heat as a quantity made this identification difficult.[48]

Just as he had for the Leibnizian and Newtonian forms of the calculus, the Scottish mathematician Colin Maclaurin showed that the Leibnizian and Newtonian forms of mechanics led to the same results. He had a preference, the same one, Newton over Leibniz, and he enlisted in the defense of Newton's measure of the force of moving bodies against Leibniz's measure, of momentum against vis viva; but he recognized that, as in the case of the Leibnizian calculus, many able foreign mechanicians used Leibnizian mechanics. To show the value of vis viva, he derived a "general Principle" starting from Huygens's principle of conservatio vis ascendentis, which he identified with conservatio vis viva, a "theorem of great use" in Leibnizian mechanics. The principle reads: If a number of perfectly elastic bodies descend, whether or not they undergo collisions, the sum of the vis viva of the bodies is the same as if all of the bodies descend freely from their respective altitudes to their several places; in collecting the sum, if any body is caused to ascend, its vis viva is subtracted.[49] Maclaurin illustrated the conservation of vis viva by solving standard Newtonian problems such as the motion of pendulums and of water issuing from a hole in the bottom of a cylinder.

Concerning the status of vis viva in mechanics, British and Continental opinions differed.[50] Maclaurin disagreed with foreign mechanicians who regarded the conservation of vis viva as a principle belonging at the head of the theory of mechanics. Because vis viva is preserved only in collisions between perfectly elastic bodies, the principle applies to "one sort of bodies only" and therefore is "not to be held a general principle or law of motion." Moreover, solutions obtained with the aid of vis viva could be obtained from Newton's universal principles, proven by the "most simple and uncontested experiments."[51] Thomas Parkinson, a Cambridge tutor in mechanics, came to a similar conclusion. Patiently reviewing the classes of experiments said to support vis viva as the proper measure of the force of moving bodies, he found them inconclusive and in no instance a disproof of the Newtonian measure, which he advocated for its universal validity.[52] George Atwood, a lecturer and Parkinson's contemporary in Cambridge, insisted that Newton's laws of motion, and these alone, belong at the head of the theory of mechanics, and that no property of motion is to be assumed other than the laws of motion, which are to be treated as axioms from which everything else in mechanics follows by deduction. The laws of motion say nothing about the conservation of vis viva or, for that matter, about the conservation of momentum. The preservation of vis viva in elastic collisions was proved by Daniel Bernoulli, which was fine, but he also regarded it as a "general law of motion," a misunderstanding of the theory of mechanics. Cases to which vis viva applies, those of constant acceleration as in falling bodies, Atwood said, are "easily deduced" from Newton's principles. The doctrine of vis viva is consistent with the "theory of motion in general," and with Newtonian momentum in particular, and it implies nothing in addition.[53] The conservation of momentum appears as a corollary to a theorem in the *Principia*, which is where it belongs, Maclaurin and Atwood believed. Had vis viva been studied by Newton, presumably its conservation would have appeared in a comparably unobtrusive place in the *Principia*, definitely not at its head. Cavendish's approach to vis viva was consistent with the general British reaction.

Beginning in the middle of the century, certain authors decided that the controversy over the proper measure of the force of moving bodies was a dispute over words. Desaguliers delayed bringing out the second volume of his course on the experimental philosophy for years because in the first volume he had promised to decide the "Question of the Force of Bodies in Motion." Because his experiments suggested that both parties were right, he announced in the second volume that what for fifty-nine years had divided the English Newtonians and French Cartesians from the German, Dutch, and Italian Leibnizians was artificial, the two sides "meaning different Things by the Word Force."[54] Maclaurin had written his views on the question years before he published them, thinking that readers might regard the dispute as just that, a quibble over words. His understanding was that "no useful conclusion in mechanics is affected by the disputes concerning the mensuration of the force of bodies in motion."[55] By the late eighteenth century, writers on mechanics could look back at the heated arguments with a degree of detachment, one likening it to the controversy "between the Jansenists and Molinists about Grace: both sides very hot, and very unintelligible."[56] There were, however, serious issues, and the controversy had implications for natural philosophy.

Vis viva was on occasion silently incorporated into British researches. Shortly before Cavendish wrote his paper on the theory of heat, John Michell made use of it in his theory of the action of gravitation on light, deducing the effect from the law of conservation of vis viva. Basing his reasoning on Newton's *Principia*, he did not use the word vis viva and took for granted its conservation in the system under consideration.[57] In time, in Britain as elsewhere, the full significance of the conservation of vis viva, or energy, as the most general principle in mechanics came to be recognized, but this could happen only after the identification of heat with motion,[58] which brings us to Cavendish.

Henry Cavendish's Researches

About Cavendish

At the start, let us have before us an image of the man: Illustration 4 shows how Cavendish looked near the end of his life. As he walks, his left arm is bent around the small of his back, while his right arm is crooked across his chest, his hand wedged inside the collar of his coat, as if applying torsion to his upper body, conveying the idea of a taut spring. Leaning slightly forward, looking dead ahead, he goes directly after whatever it is he desires. He is not heavy, but beginning to thicken, his rumpled coat hitching over his hips. Emerging from beneath a three-cornered hat, hanging down his back, is a knocker-tailed wig, long out of date. His appearance is decidedly unprepossessing. The sketch from which we learn these particulars was made surreptitiously, since he steadfastly refused to sit for portraits.

Cavendish was born in 1731 in Nice,[1] where his parents went because of his mother's health, which continued to decline. Shortly after giving birth to a second son, she died. Henry was two, and his father never remarried. At eleven, he was sent to a progressive school near London, Hackney Academy, from where, in 1749, he proceeded to St. Peter's College, Cambridge University. As was customary for Fellow Commoners, after about three years, he left without taking a degree. Thereafter he lived with his father in Great Marlborough Street, London, where, not having to earn a living, he devoted himself to scientific pursuits. At about the time his father died in 1783, he resettled, moving into a house on Russell Square and into another outside London, on Clapham Common. The first he converted into a scientific library, open to qualified users, the second into a workshop, laboratory, and observatory for his own exclusive use. In these places of science, he spent his days. He never married.

Illustration 4. Henry Cavendish: graphite and gray wash sketch by William Alexander. Reproduced by permission of the Trustees of the British Museum.

Apart from his science, the most important fact about Cavendish was his position in society. Descended from the Duke of Devonshire on his father's side and the Duke of Kent on his mother's, he was known as the Honourable Henry Cavendish. Proud of his family name, its station, and its politics, he was an aristocrat at a time when the aristocracy was in high tide. He accepted his preeminence in society as he did the laws of nature. The next most important fact about him was his father, Lord Charles Cavendish. An accomplished experimentalist and respected administrator of the Royal Society of London, his father instructed him in science and saw to it that he had an appropriate formal education, after which he introduced him to the scientific society of London. Like his father, he became a prominent member and administrator of the Royal Society. Father and son were motivated by an ethic that went with their class, the duty of public service, which they performed in an unusual venue, the halls of science.

Late in life, Cavendish was as well known for his wealth as he was for his science. He owed his wealth to the Cavendish family, whom we meet in the following scene. It was not uncommon for untold numbers of guests to arrive at Devonshire House, the Picadilly mansion belonging to Cavendish's cousin, the Duke of Devonshire. Deposited by their carriages behind the high wall at the street that separated the Cavendishes from the mob were lords and ladies, cabinet ministers, generals, diplomats, solid men of learning, and various Cavendishes in town for the season. The gay crowd circulated through the gorgeous William Kent reception rooms, passing beneath Old Masters displayed on luminous, silk-covered walls, and spilling out into the spacious gardens behind. Henry Cavendish, however, was not there. The only times he set foot in this palace were the rare christenings or the scientific conversations he had with the Duchess of Devonshire, the queen of London fashion who also had a keen if unfocused curiosity about the universe. Cavendish honored the achievements of his family, but he had no desire to emulate its display. He spent little time counting out his fortune, which in any case came late in life and had no effect on his tastes, habits, and outlook.

Not in the great houses of the aristocracy but in a solid brick house on Clapham Common was where Cavendish was usually to be found in later years. This house was set apart not by grandeur but by the towering ship's mast in the back, a mount for aerial telescopes, and upon closer inspection, by the thermometers and other instruments stuck all around it, and inside by its laboratory for a drawing room, forge for an adjacent room, and observatory for another room. Cavendish's premises gave the largely correct impression that he made little distinction between his personal affairs and natural philosophy.

Beginning at age thirty-three, Cavendish submitted papers to the Royal Society to be read at meetings and then published in its journal. They were not many, fewer than twenty over a career of nearly fifty years, but some were choice. His most important papers were on pneumatic chemistry, electricity, and heat, the three leading fields of experimental research in the second half of the eighteenth century. The extensive scientific manuscripts recovered after his death reveal a wealth of experimental, observational, and mathematical studies belonging to all parts of natural philosophy. The skill and foresight he showed in these private studies have fascinated scientists ever since, as has his seeming indifference to their publication. His contribution to science was large, first for his example of exacting procedure, second for his labors in the Royal Society on behalf of the enterprise of science. He remained active in science in both respects to the very end of his long life, in 1810.

It used to be thought that Cavendish was an experimental chemist and physicist of an extremely limited sort, one who was never caught without an instrument in hand, a compulsive quantifier of whatever chanced to lie in front of him. A scholar who does not accept that characterization of him lays to rest the ghost of Cavendish, the measuring machine: "[T]his most creative scientist of the eighteenth century was described as without imagination, the unity of his work was lost in apparent diversity, and his desire for completeness and for quantitative verification was transmuted into whim and a compulsion for weight, number, and measure."[2] Cavendish was the first of Newton's countrymen to possess a similar, impressive combination of experimental, mathematical, and theoretical skills.

To get on with its work, the Royal Society shunned political, religious, and other polemical distractions, a policy that fit perfectly with Cavendish's natural inclinations. Although he belonged to a politically active family, he rarely mentioned politics, and when he did, it made news.[3] The one occasion on which he is known to have engaged in political activity was an internal disruption of the Royal Society, where the threat to open scientific exchange was sufficiently great for him to overcome his reluctance to take part in controversy.[4] Never speaking of religion, in the matter of faith, he was said to be "nothing at all."[5] Never attending a place of worship, his only known connection with the church of his parish was his mention of it at a scientific club: Part of the church, he told the members, was eaten "thro' by the insects . . . working their way out."[6] Scientific research and reading being his sole occupation and amusement in life, his society of choice was scientific men, whom he sought out at coffeehouses, instrument shops, and, above all, the Royal Society.

Meetings of the Royal Society were held in Sommerset House, a monumental stone structure beside the Thames. Every Thursday Cavendish passed beneath British Arms supported by the Fame and Genius of England, beneath a fitting example, a bust of Newton, and then entering a small room, he took his place on one of the benches for the members. An isle dividing the benches led to an elevated chair where the President sat, in front of which stood a desk for the two Secretaries, one of whom read aloud from scientific papers recently communicated to the Society. From paintings hung high on the walls, Newton and other illustrious past members watched over the serious scientific proceedings. Seated there among his colleagues, listening intently to the papers, on occasion one of his own, Cavendish was in his element.

Typically, Cavendish began his day by checking his earth-magnetic and other instruments in the garden. Then moving indoors to the laboratory, he picked up where he had left off the day before, performing experiments and doing what writing went with them. Then at an appointed hour in the afternoon, he took a predetermined solitary walk around the fields behind his house. In the evening he read in his scientific books and journals. Unless he journeyed into the city, that was his day, as we imagine it. If his banker called upon him without warning, he was upset, having arranged his life precisely to discourage such interruptions. To make certain that even his servants did not disturb him, he communicated with them by writing and built a separate staircase for their use. Repetitive and uneventful, his outward existence was one of nearly perfect serenity.

Our picture is missing an important detail. Every Sunday afternoon, Cavendish attended a soiree at the house of the president of the Royal Society. Strangers were usually present, and from firsthand accounts of these affairs, we learn that Cavendish dreaded the strangers' gaze. He was, as it were, endowed with two sets of eyes, one set safely behind lids that could close out the world, the other set free to wander, to regard himself from all directions, inquisitionally and pitilessly, as he supposed strangers did. Strangers no doubt did stare at this celebrated man of science, and what they saw was remarkable. Acutely self-conscious, embarrassed, and semingly gripped with fear, Cavendish shrank from their presence, lost control of his voice, fell silent, and became invisible. His oddities in public had a counterpart in science, where his multiple sets of eyes proved an asset. What gave rise to anecdotes about his behavior was in the privacy of his laboratory a profound circumspection. Paralleling a critical, objective self-scrutiny was a masterful control over experimental error, ensuring that once he committed himself on a scientific question, he rarely had to retract. The relentless hounding of error to get at the truth of nature lent this strange man a very considerable dignity. As truth was understood in science, Cavendish embodied it.

Whether he was practicing natural philosophy or engaging in any other activity, Cavendish conducted himself the same way. He was described by a close colleague as having a "truly philosophical simplicity of manners" in "private life."[7] Simplicity was his hallmark. His handwriting was clear, bold, and unembellished. He insisted on no form of address for himself. His clothes were plain, his fare was plain, his house was plain, and his scientific instruments and apparatus were plain. These details were commented on by persons who knew him, who also commented on his integrity, candor, and love of truth. If this characterization sounds fawning, these same traits ideally underlay the practice of natural philosophy.

Natural philosophy had many uses. Uplifting to the parson composing a sermon about Creation, entertaining to the gentleman making artificial lightning in his parlor, and instructive to the artisan witnessing demonstrations of the laws of mechanics, it offered a way of life to a person like Cavendish. This may say as much about natural philosophy in the eighteenth century as it does about his life. "Natural philosophy" acquires yet another meaning.

Principles of Mechanics

Mechanics was, among other things, a repository of concepts and laws for making causal theories in other departments of natural philosophy. Because electric sparks and warmth to the touch were not like comets moving across the sky or balls rolling down inclined planes, bodies for which the rules of mechanics had been devised, Cavendish had to rely on analogy to compare electricity and heat with mechanics.

The science of mechanics underwent vigorous development in the eighteenth century; its most important contributors; were, as it happened, European, especially the Swiss and the French. Leonhard Euler, who has been called the "dominating theoretical physicist" of the century, devoted his career to developing and supplementing Newton's work, placing rational mechanics on the foundations it has retained to this day. It was he who first stated the laws of motion as differential equations in rectangular coordinates, known today as Newton's equations of motion. Euler, James Bernoulli, and others developed the great body of mechanics of deformable bodies. The most productive period in rational mechanics began in the 1730s and ended at mid century, with opportune timing for Cavendish, who was just then finishing his studies in Cambridge University. When he began his scientific work, there was much that was new in mechanics, much that was not in the *Principia*.[8]

Fortunately for us, Cavendish wrote down his thoughts on mechanics. Although his electrical researches expanded to the point where he evidently considered bringing them out in the form of a book, the only branch of natural philosophy on which he set out to write a systematic treatise from the start was mechanics. Our evidence, "Plan of a Treatise on Mechanicks," can be roughly dated, because it contains a reference to a paper on mechanics by Hugh Hamilton in 1763. Because he was thirty-two in the earliest year that the "Plan" could have been written, we may take it as a guide to his mature thinking on mechanics.

As far as the "Plan" informs us, there are two parts to the intended treatise. The first part, statics, the "doctrine of pressures & the mechanic powers," begins with the rule of composition and resolution of forces and a fundamental proposition about the lever, "very well proved by Maclaurin," from which the other mechanic powers such as the pulley are deduced. Independent of the "properties of matter," which can "be proved only by experiment," the doctrine of pressures and the mechanic powers is capable of "demonstration independently of experiment as well as any other mathematical truths."[9] Newton's treatment of statics was limited to a single corollary to his laws of motion, the weakest part of the *Principia*;[10] statics offered Cavendish an opportunity to improve upon the original.

The second part of Cavendish's planned treatise treats dynamics, the "theory of motion." The laws of motion, two in number, not Newton's three, are mathematical statements about uniform and accelerated motion, the "same as S[r] I[saac] N[ewton']s 2 first laws of motion & comprehend all we know of the properties of matter with regard to motion." The equality of action and reaction "is called by S[r] I[saac] N[ewton] the 3rd law of motion but improperly," because "this axiom is merely a property of the doctrine of pressures."[11] His heterodoxy on this point had support.[12] Reversing Newton's reasoning, Cavendish deduced the equality of action and reaction from the lever.

The plan of the treatise on mechanics does not go beyond the laws of motion, to which point the discussion is elementary. That raises the question of what Cavendish had in mind. Had he been in need of income or had he been a teacher, he would likely have written a popularization or a textbook, but he was

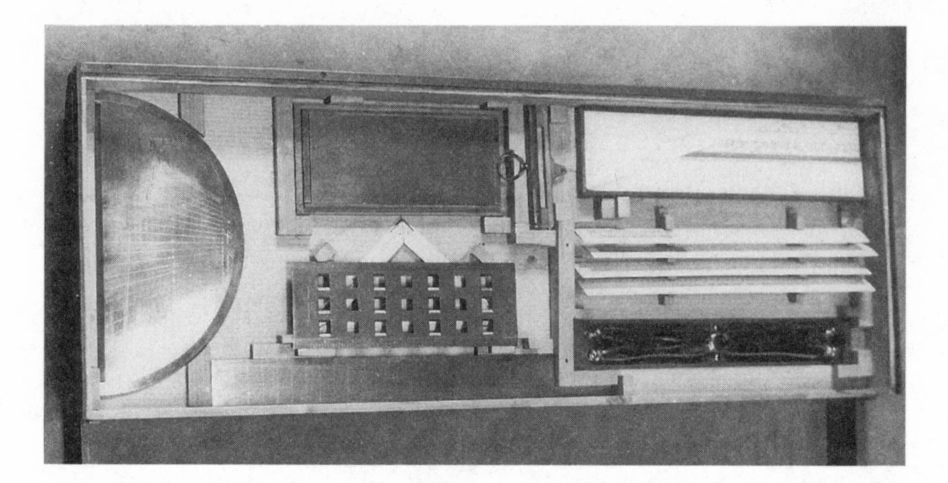

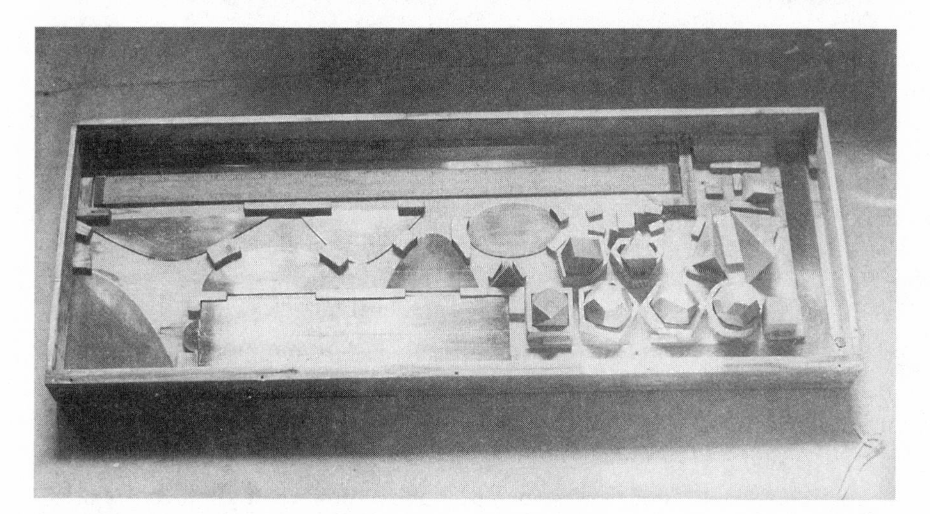

Illustration 5. Mathematical instruments. As well as in their laboratories and observatories, natural philosophers used instruments at their writing desks. The two drawers shown here, from a cabinet belonging to Cavendish, contain scales and rulers made of brass and wood, an ivory triangle, boxwood regular solids, a brass globe map projection, and so on. With these instruments, inscribed with the names of well-known instrument makers such as Jesse Ramsden and Jonathan Sisson, Cavendish made drawings for his papers, several of which are reproduced in this book. Photograph by the author. Devonshire Collections, Chatsworth. Reproduced by permission of the Chatsworth Settlement Trustees.

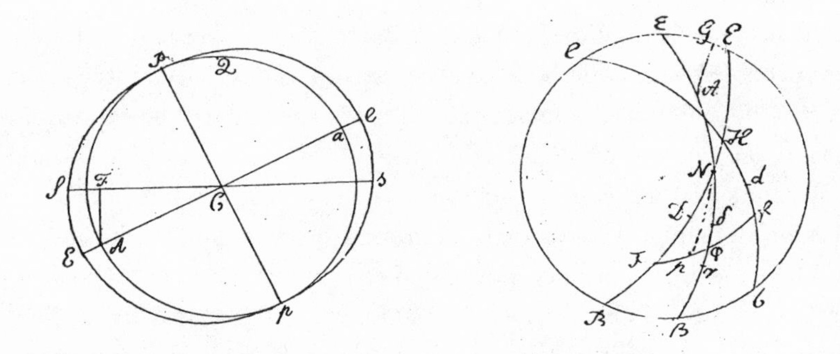

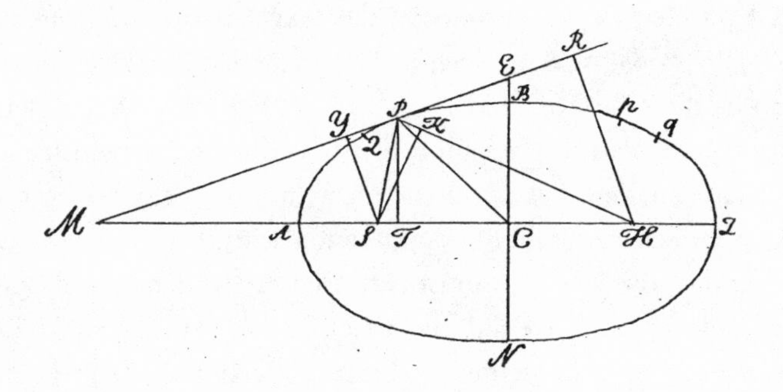

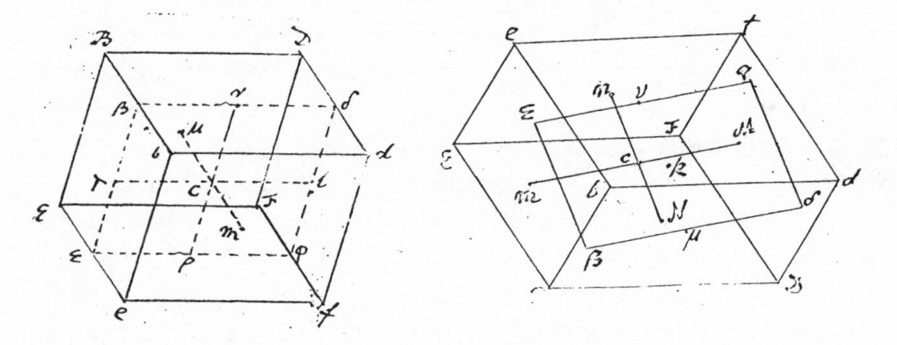

Illustration 6. Partial theories of mechanics. The figures shown are auxiliary constructions drawn by Cavendish in connection with various mechanical investigations: Fig. 1, upper left, precession of the equinoxes; Fig. 2, upper right, path of a comet; Fig. 3, middle, path of a planet in a resisting medium; Figs. 4 and 5, lower left and right, motion of a solid of revolution from centrifugal force. Cavendish Mss V(b), 8; VIII, 6, 9, 43. Reproduced by permission of the Chatsworth Settlement Trustees.

neither. Because at the time he planned his treatise on mechanics he was otherwise occupied with entirely original researches in heat, electricity, and chemistry, it is possible that he intended it as a scientific contribution. From other papers he left, we can make a case for it.

Watermarks on Cavendish's stationery suggest that a number of his many studies in mechanics were carried out around the time he planned his treatise. They include work on problems discussed in part 1 of this book, such as Vince's theory of friction, Robins's theory of gunnery, and the loss of motion of a projectile owing to the resistance of air.[13] Cavendish performed experiments on the effects of resistance such as on the motion of water through a tube.[14] He investigated problems that lay outside the range of the *Principia* such as the spinning top;[15] problems that Newton had treated impressively but inconclusively, such as the efflux of fluid from a vessel[16] and the shape of the earth;[17] and problems such as the precession of the equinoxes, in which Newton was believed to have made a flat-out error.[18] On the subject of Newton's most extended experimental researches in conjunction with theory, pendulum motion, Cavendish, too, carried out his most extended experimental and mathematical researches in mechanics.[19] He did mathematical and experimental work in areas of greatest achievement in eighteenth-century mechanics, in theories of solids, elastic vibrations, and fluids.[20] To Newton's largely erroneous theories of fluids,[21] Cavendish gave repeated and critical attention: The motion of water waves "as S^{r} I[saac] N[ewton] has described it . . . is not the case"; "S^{r} I[saac] N[ewton']s demonstration concerning sound . . . does not agree with experiment."[22] Cavendish's extensive researches testify to his competence in all parts of mechanics.

For an unspecified reason, but perhaps for his treatise, Cavendish made a study of the slowing of the earth's rotation caused by the friction of the tides, in which he deduced the "loss of force by friction" from the "visible" and "invisible" vis viva of the moon, earth, and water.[23] This is the same terminology he used in his one extended study of vis viva in physics, "Heat." We now turn to his interest in vis viva.

Theory of Motion

Before discussing Cavendish's work on the theory of motion, let us learn something of Daniel Bernoulli's. Of Continental physicists and mathematicians, Bernoulli was probably the closest to Cavendish in the general direction of his science, although his most important work was mathematical whereas Cavendish's was experimental. Had Cavendish's "Heat" had a precedent, most likely Bernoulli would have been its author. The first to join Newton's physics to Leibniz's calculus, Bernoulli fought Newton's battles on the Continent, while profiting from the powerful methods of mechanics and mathematics originating there. A founder of mathematical physics, he is described by a biographer as "first and foremost a physicist, using mathematics primarily as a means of exploring reality as it was revealed through experimentation."[24] This trait comes through in his *Hydrodynamica* in 1738, which he characterized as "Physical rather than Mathematical"; lacking in "mathematical rigor," his treatment of fluid mechanics relied on a "physical hypothesis," and throughout this work, he compared his and others' experiments with his theory.[25]

It was Bernoulli who introduced into physics the principle of conservation of vis viva and virtual work, a limited form of our principle of conservation of energy.[26] The theory in *Hydrodynamica* is based on the equality between the descent and the ascent of the center of gravity of a system of particles,[27] from which the principle of "conservation of live forces" follows at once. If in collisions between inelastic bodies, vis viva seems to be lost, the universality of the principle is not compromised, since a portion of vis viva "remains impressed in the certain fine material to which it has transferred." Because Bernoulli held a mechanical view of heat—"the greater the heat is, the more violent is the motion of all the particles"—he might reasonably have gone on to equate heat with vis viva equivalent to the gravitational potential of a raised body, but he did not.[28] He later generalized the conservation law from a system moving under a constant gravitational force to a system in which bodies interact according to any law of mutual "gravitation," or central force.[29] We return to Bernoulli when we explore Cavendish's derivation of the conservation law.

For us, Cavendish's most significant mechanical writings were two studies of vis viva, "Articles Relating to Theory of Motion." The labeling is curious, because he ordinarily referred to his studies as "papers," and "papers" were read before the Royal Society and published in the *Philosophical Transactions.* He may have intended his "articles" on the theory of motion as contributions to a publication other than the journal of the Royal Society, although for various reasons that seems unlikely. A more general meaning of "article" is an independent part of a whole, which is how Cavendish referred to numbered discussions in "Heat." His articles on the theory of motion may well have been intended as independent parts of his treatise on mechanics. Indeed, if he had plans for them, we can think of no other place where they might go.

One of the articles is a brief clarification of the "famous controversy about the force of bodies in motion."[30] Sensitive to the use and misuse of words, he agreed with the position that the two sides in the controversy meant something different by the expression "force of bodies in motion." In view of his subsequent use of vis viva, his clarification of the controversy is significant. Freed from the confining polemics, he could fairly take advantage of what vis viva had to offer him.

The editors of Cavendish's *Scientific Papers* omitted the brief article on the controversy but published the substantial second article, to which they gave the title "Remarks on the Theory of Motion."[31] This article begins with a summary of the first, setting aside the controversy, then goes on to develop the theory of vis viva and give examples of its use. It contains neither dates nor citations to publications, but its discussion of heat offers a clue as to when it was written. The heats accompanying fermentation, dissolution, and burning, and also the cold produced by certain mixtures, had been recogized as difficulties for the view of heat as motion,[32] and in "Remarks" Cavendish said that he did not know how to resolve them. His list of difficulties did not include specific and latent heats, a significant omission; Black regarded the discovery of specific heat as "very unfavorable" to the Newtonian view of heat,[33] and the same was said of the discovery of latent heat.[34] Because elsewhere Cavendish acknowledged these difficulties, either "Remarks" came very early in his work, before his experimental investigation of specific and latent heats, or it came after he had resolved the difficulties; I suspect that "Remarks" came early.

We are not told where Cavendish first learned about vis viva, whether through Continental writings such as Bernoulli's, or through less complete and appreciative accounts in English such as Maclaurin's, or through hostile polemics. There is only one citation in "Remarks," that to a proposition in the *Principia*. We do know that by the time of this article, he was thoroughly familiar with the concept and prepared to extend its applications throughout natural philosophy.

For most questions arising in "Philosophical enquiries," Cavendish wrote in "Remarks" that the usual and most convenient way of "computing the force of bodies in motion" was Newton's momentum, the measure of a pressure acting on a body over a given time. The alternative measure, that of a pressure acting on a body over a given space, was usually reserved for solving problems concerned with machines for mechanical purposes, but it was "also very often of use" in philosophical inquiries. For this measure, Cavendish used the terms "mechanical force or *mechanical momentum*" instead of "vis viva."[35] By his

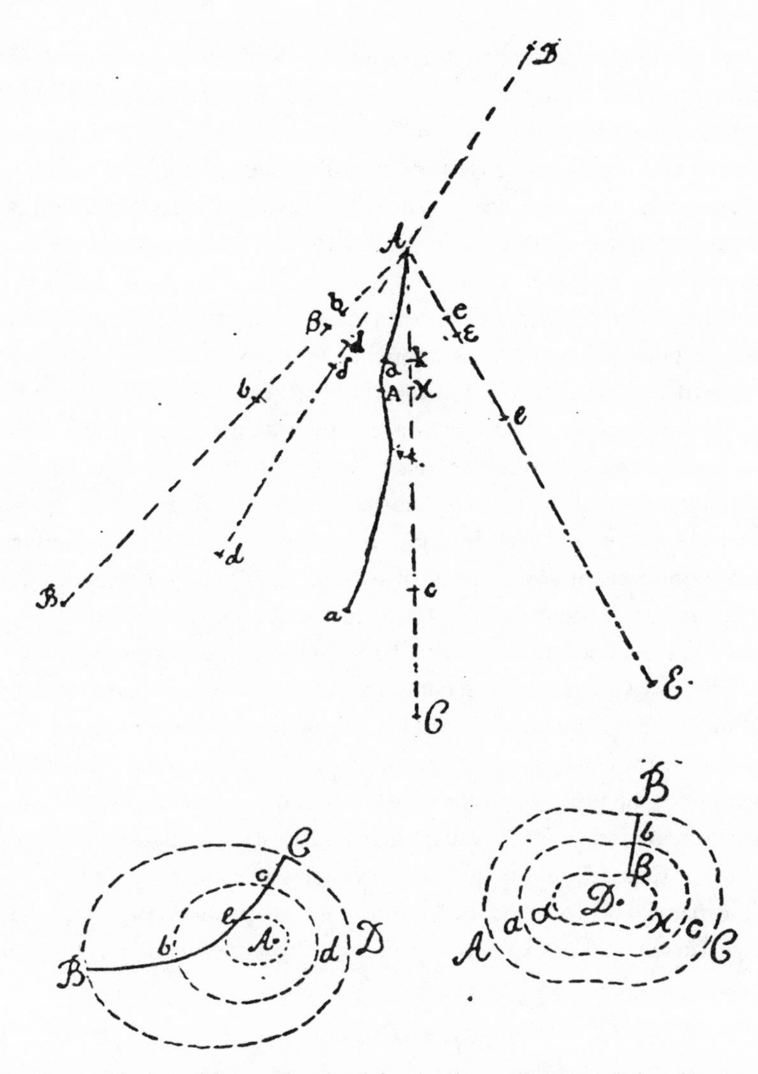

Illustration 7. Theory of force. The three figures drawn by Cavendish refer to central attracting and repelling forces. By assuming that the forces are "always equal at equal distances however unequal at unequal distances," he derived the law of conservation of actual and potential mechanical momentum, or vis viva. Figure 1 (top) accompanies an analysis of the vis viva of body A under the action of forces centering on B, C, D, and E that satisfy the assumption. Figures 2 (bottom left) show curves of constant central forces that do not satisfy the assumption. Figure 3 (bottom right) is a special case: because the distances between the curves are constant, Cavendish argued, the conservation law still holds. We see an (unsuccessful) attempt, Cavendish's editor observes, to formulate the modern concept of equipotential curves. Cavendish, *Scientific Papers* 2:425. Cavendish Mss Misc. Reproduced by permission of the Chatsworth Settlement Trustees.

choice of wording, referring to both ways of computing the force of moving bodies as species of "momentum," he acted on his conviction that they were alternative, equally valid measures of force.

In "Remarks," to demonstrate the value of vis viva, Cavendish derived a law of conservation of vis viva for a system of moving bodies interacting by any attracting and repelling forces, provided that the forces "are always equal at equal distances" from their centers: If no force is lost by friction and inelastic collisions, he wrote, the sum of mechanical momenta and "additional momenta," or potential mechanical momenta, is always the same. In corollaries, he discussed five kinds of phenomena to which the law applies. The first is a system of any number of moving bodies such as a myriad of particles. The second is the motions of the particles of bodies constituting heat. The third is the interactions of the particles of light and the particles of bodies. The fourth is the motions of the elastic fluid of the atmosphere constituting sound. The fifth is the motions of any number of bodies connected by springs or by rigid rods with frictionless joints. The last corollary is followed by a succinct statement of the conservation law, which brings the paper to a close.

The editor of Cavendish's mechanical manuscripts, Joseph Larmor, delivered the following judgment on Cavendish's law of conservation of actual and potential vis viva and his illustrations of it: "This surely is the earliest precise enunciation of the principle of the conservation of energy, kinetic and potential, including enumeration of the causes that lead to its degradation, which on the principles of Cor. 2 would be into heat of precisely equivalent amount."[36] Cavendish expanded Corollary 2 to become the independent paper "Heat," and although Larmor did not have this paper, it fully supports his judgment above.

Importance of Heat

Having in an earlier chapter examined Newton's view of heat in the operations of nature, let us now look at views of heat a century later, in Cavendish's time. We first consider heat as an industrial agent, then as a subject of natural philosophy.

The British industrial revolution has been dated variously, beginning in the year 1760, according to one assessment, the same year that Cavendish entered the Royal Society, a noteworthy coincidence. Alongside his first interest, natural philosophy, he developed a keen interest in industry. For several years in the mid 1780s he and his associate, Secretary of the Royal Society Charles Blagden, made extensive journeys throughout Britain. By this time an extraordinary landscape of furnaces, mills, and machinery had come into being, and Cavendish ventured into it with all the curiosity he brought to his studies in natural philosophy. The same years saw his researches on heat, including his mechanical theory of heat.

Cavendish and Blagden's first journey took them into Wales, where they witnessed a wide range of industrial processes: quarrying, cloth manufacturing, dying, coal mining, coal-tar manufacturing, lime kilning, coke making, copper casting, brass drawing, and iron making. They saw iron furnaces standing as high as forty-five feet. They saw forges with their intense heat and violent fireworks, and they saw burning coal pits. They saw heat and fire everywhere.

They stopped at the Soho Works outside Birmingham to visit James Watt, who had invented the separate condenser for the steam engine in the 1760s, and who had just made another major improvement that converted the linear motion of the piston drive to rotary motion, highly useful in mills. They saw Watt's improvement and also a furnace he contrived for burning smoke, which he intended to apply to the steam engine. Cavendish made drawings of Watt's rotative mechanism and smoke-burning furnace.[37] On their journey, he came across Watt's steam engines in use.[38] That fall Watt came to London, to Albion Mills at Black Friars Bridge, for the installation of his new smoke-burning furnaces. We can imagine that Cavendish was on hand; we know he went there with John Smeaton to inspect plans for a steam engine.[39] Except for the difference of scale, there was an unmistakable similarity between industrial Britain and Cavendish's laboratory, where with his burners and pots he investigated the laws of heat, and with his small furnaces he examined the specimens he brought home from the great furnaces he had witnessed in operation.

As it was so patently a mover of industry, heat was a mover of nature, at times a mighty force. If electricity in the form of lightning was capable of destroying a house, fire was capable of destroying a city, as in the great London fire, or the Lisbon earthquake, thought to have been caused by subterranean fire.[40] Josiah Wedgwood, whose ceramic manufacture demanded high temperatures, performed experiments on that "grand and universal agent," heat. With a thermometer based on the shrinkage of pieces of clay, he extended the scale of heat from the boiling point of water to over thirty thousand degrees Fahrenheit. The freezing point of water to the vital heat of humans took up only one five-hundredth of Wedgwood's scale.[41] As the earth seemed increasingly insignificant on the scale of the universe as revealed by astronomy, the range

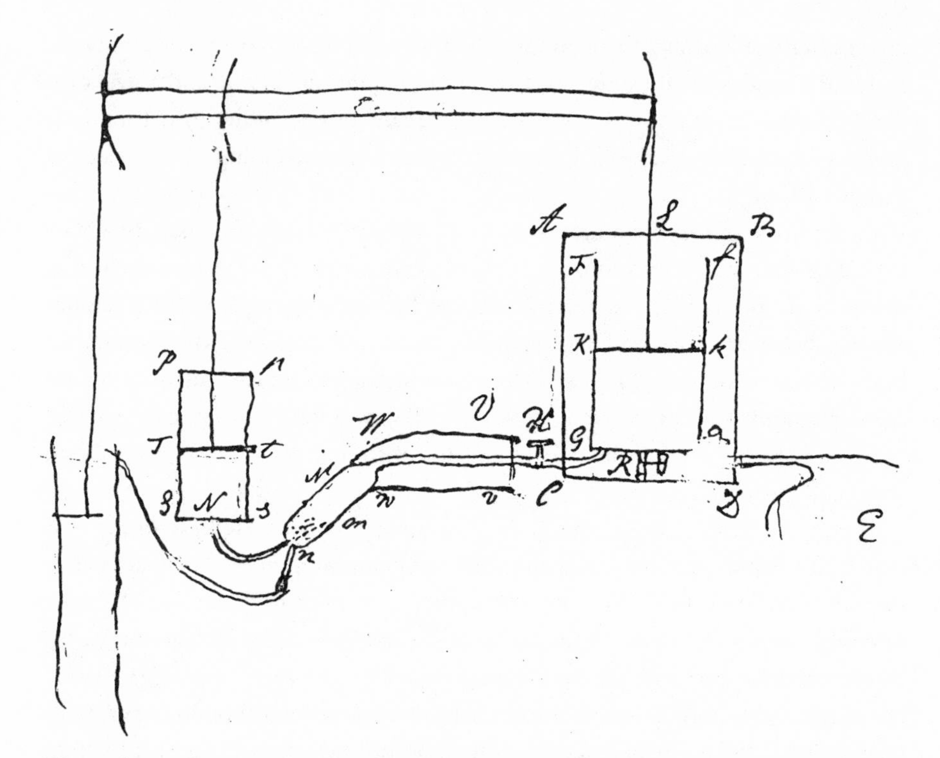

Illustration 8. Steam engine. Drawing by Cavendish. FfGg is the working cylinder, kept hot by a steam jacket. Steam is condensed in the vessel GMmN, cooled by water running in at n. The cylinder on the left works in the manner of an air pump to preserve a vacuum in the condenser. Cavendish compared this, James Watt's, engine with the "common" engine. In the latter, which did not have a separate condenser, the working cylinder had to be reheated every cycle by a fresh charge of steam; because of the resulting condensation, the steam lost considerable volume and pressure. Cavendish Mss Misc. Reproduced by permission of the Chatsworth Settlement Trustees.

of heat normally experienced on earth seemed insignificant on Wedgwood's scale. Cavendish paid a visit to Wedgwood on his journeys.

Rumford, a principal investigator of heat, ranked heat with gravity as a natural power: "[T]he effects produced in the world by the agency of Heat are probably *just as extensive*, and quite as important, as those which are owing to the tendency of the particles of matter towards each other," and "its operations are, in all cases, determined by laws equally immutable."[42] Of equal significance as gravity, heat awaited its Newton, the natural philosopher who would discover its laws, develop its theory, and erect a thermal system of the world to stand beside the gravitational. Heat, Black told his students, "is certainly the chief material principle of activity in nature," and if it were to be removed, "a total stop would be put to all the operations of nature."[43] Black's student Cleghorn expressed his opinion of the importance of heat in nature in words almost identical to Black's. In the absence of heat, he said, "Nature would sink into chaos," the bleakest prospect imaginable to the eighteenth-century mind. Of heat, Cleghorn concluded, "nothing will seem more deserving of the attention of philosophers."[44] Cavendish agreed.

Rules of Heat

Cavendish began experimenting on heat around the same time as, probably slightly later than, Black. His earliest recorded experiments include an extended series of studies on specific and latent heats. The few dates in the minutes of the experiments are in order, and the sequence of experiments follows a natural progression of questions and answers.[45] The earliest date, 5 February 1765, occurs near the end, suggesting that Cavendish began his experiments on heat no later than 1764. The minutes convey the feel of experimental research leading to unanticipated results and, as we will see, to a theory of heat.

The occasion for the experiments is uncertain, but in a broad sense, their sources are evident. First, Cavendish's tutor in science, his father, Lord Charles, took a special interest in thermometry and in the change of state of matter. Second, just as Cavendish began his career, in the 1760s, the experimental field of heat emerged as a quantitative science. This, as we know, was the kind of science in which Cavendish excelled.

As opportunity allowed, Cavendish kept abreast of developments in heat. In many respects, London was the scientific center of Britain, but the most important researches on heat were done not there but in faraway Scotland. In the minutes of his experiments, Cavendish mentioned only one name, "Martin," the Scot George Martine, who investigated rates of heating and cooling.[46] In a paper based on these minutes, he mentioned three other names, all in connection with latent heat. One was a French physicist, whose work was some twenty-five years old.[47] The other two were the Scottish chemists who introduced the subject of specific and latent heats, William Cullen and Joseph Black.

Cavendish may have learned of Cullen's experiments on cooling by evaporation as early as 1756, from the Edinburgh *Essays* or from a reference to that publication in the *Philosophical Transactions* the next year.[48] As a regular at the meetings of the Royal Society and at the dinners of its club, Cavendish probably heard of work on heat in Scotland from Scottish guests passing through town or from others who had heard about it. It may have been from his friend John Hadley that he learned of Cullen's work and, perhaps at the same time, of Black's. The evidence is a letter by Benjamin Franklin in 1762, in which he spoke of witnessing Hadley's repetition of one of Cullen's experiments on cooling.[49]

It is possible that Cavendish took up researches on latent heats and heat capacities after hearing of similar work in Scotland, but it is also possible that he came to the subject independently by following more or less the same route that Black took. A likely common source was Boerhaave's text on chemistry, which Black found a helpful guide and which was recommended reading at Cambridge when Cavendish was there.[50] In it Cavendish would have read about Fahrenheit's observations of changes of state attended by heats that do not register on the thermometer, an instance of latent heats, and also about Fahrenheit's observations of the different heating effects of mercury and water and his conclusion that the two substances have different heat capacities. Like Black, Cavendish began his experiments on heat by investigating the thermometer, determining that mercury expands in proportion to the heat. The obvious precedent here was Brook Taylor's experimental study of the a linseed-oil thermometer.[51]

To carry out experiments, Cavendish needed little in the way of equipment. Besides the principal instrument, the thermometer, he had a watch, scales, lamps for heating, and tin vessels. Like his predecessors Fahrenheit, Taylor, and Black, he began with the simplest instance of heat exchange, a mixture of hot and cold water. Recording three readings three minutes apart, he determined the rate of heat loss to the surroundings, and he did a separate experiment to determine the heating effect of the apparatus. "It seems reasonable to suppose," he said, "that on mixing hot and cold water the quantity of heat in the liquors taken together should be the same after the mixing as before; or that the hot water should communicate as much heat to the cold water as it lost itself."[52] By experiment, he showed that the "true heat" of the mixture of hot and cold water is the weighted mean of the temperatures of the portions before mixing. Once confirmed, this supposition became Cavendish's first general fact, the rule of conservation of heat. In the minutes of his experiments, he could then write, "[T]herefore the heat of the mixture ought to be by theory. . . ."[53] The conser-

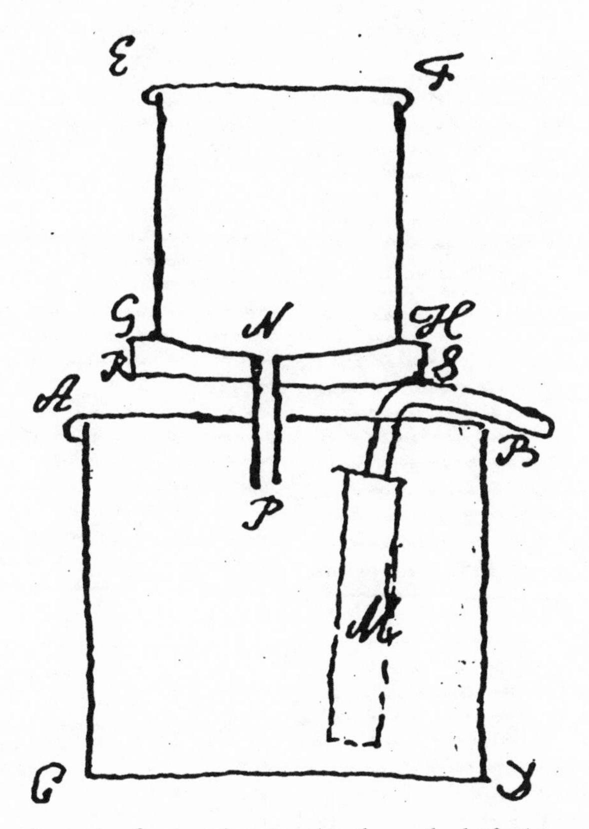

Illustration 9. Apparatus for experiments using the method of mixtures. Drawn by Cavendish. From the cylindrical funnel on top, hot water is added to cold water in the pan below. Other substances can be mixed in the same way. M is a stirrer. Cavendish Mss Misc. Reproduced by permission of the Chatsworth Settlement Trustees.

vation rule together with his finding about thermometers laid the foundation for a quantitative investigation of the entire subject. We see from Cavendish's first experiments that he, like Black, grasped the distinction between, and the need for, two basic concepts in the quantitative study of heat: quantity of heat and intensity of heat, or heat by the thermometer. This basic understanding became clear only in the 1760s.

Next Cavendish tried a more complicated heat exchange, mixing water at one temperature with other substances at other temperatures. The result was not what his readers expected: "One would naturally imagine that if cold ☿ or any other substance is added to hot water the heat of the mixture would be the same as if an equal quantity of water of the same degree of heat had been

P.48

Exper. on heat of mixture of hot & cold water

The cold water was put in large tin pan with cover the hot water was put in a tin funnel the pipe of which passed through a hole in cover of tin pan this funnel was furnished with a cover with a stirrer & a hole to let in a thermom. the pipe of the funnel was stopped up with cork which at the proper time was drawn out by a piece of wire passing thro the cover So soon as the hot water was let in the funnel was taken away & the ball of the therm. let in by the hole which the pipe passed through The cover of the tin pan was also furnished with stirrer A piece of board was placed under funnel to prevent its heating cold water which board was kept from touching cover by 3 bits of wood about $\frac{1}{4}$ inch thick & in all the

P.49

exper. except the first the funnel was surrounded with pasteboard case stuffed with wool

Exp. 1

at 5.55 cold water at $48\frac{2}{3}$
at 6.12.10 water in funnel at – 197
12.50 ——— 196
13.25 – – – – – 195
at 16.0 mixture at — 99 –
20.30 — — — — $98\frac{1}{2}$ –
weight cold water = 249.8 oz
—— mixture = 383.8
hot water = 134

cold water increased in heat about 1° in 50' therefore its heat at ~~…~~ 49 the time that hot water was put in was 49°

The hot water took up near 30'' to run intirely out of funnel but ran almost out in 15 or 20'' Therefore ~~heat of hot water~~ supposing it to take up 15'' heat of hot water is 194,6 mixture cooled $\frac{1}{4}$° between 13:40'' & 16.0'' therefore true heat mixt. = 99.2

P.50

The weight of the tin pan was 31.5.0 that part of the stirrer ~~under water~~ within the pan was 2.6.0 the sum is 33,5 oz & the effect thereof ~~…~~ supposing them to be heated equally with water is 3,94 oz as iron is equiv. to about $\frac{1}{8\frac{1}{2}}$ its weight of water the cover weighed 9,6 oz & therefore its effect supposing it to be heated $\frac{1}{2}$ as much as water is equal to ,56 of water so that $4\frac{1}{2}$ oz are to be added to weight of water on account of pan &c Therefore the heat of the mixture by computation should be 99,24

N.B ~~the ther~~ in trying the heat of hot water & mixt. so little of the scale was out of the pan that no sensible error could arise from thence

P.51

2d exper. in same manner

at 6.13.0 cold water at 47° –
at 6.25.20 water in fun. at 199
26.0 – – – – – 198
26.50 – – – – 197
27.40 – – – – 196
at 30.20 mixt. at – – $97\frac{3}{4}$
weight of cold water = 249,8
mixture ——— 382,5
hot water ——— 132,7

computing as in former exper
true heat cold water = 47,2
hot water = 195,7
mixture = 98.

heat mixt. by computation should be 98°,1

Illustration 10. Minutes of experiments on mixtures: the first two of a series of experiments by Cavendish on the heat of a mixture of hot and cold water. The heating effect of the parts of the apparatus, which is expressed in terms of the "equivalent" weight of water, is taken into account, and the "true" heat of the mixture is adjusted for the cooling that occurs during the time of the experiment. Cavendish Mss III(a), 9: 48–50. Reproduced by permission of the Chatsworth Settlement Trustees.

added; or, in other words, that all bodies heat and cool each other when mixed together equally in proportion to their weights. The following experiment, however, will show that this is very far from being the case."[54] Using the same method but varying the apparatus, he alternately mixed hot water, mercury, and alcohol with cold mercury, alcohol, and a number of other substances, taken from his shelves of chemical reagents: oil of vitriol, solution of pearl ashes, sand, iron filings, shot, powdered glass, marble, charcoal, brimstone, and Newcastle coal. From these experiments, Cavendish arrived at his next general fact, another law: "It should seem, therefore, to be a constant rule that when the effects of any 2 bodies in cooling one substance are found to bear a certain proportion to each other that their effects in heating or cooling any other substance will bear the same proportion to each other." Having experimentally established a rule of nature, Cavendish next assigned to it a physical cause: The "true explanation," he said, is that "it requires a greater quantity of heat to raise the heat of some bodies a given number of degrees by the thermometer than it does to raise other bodies the same number of degrees."[55] With a rule and an explanation, he had a theory of specific heats.

With his theory as a guide, Cavendish carried out further experiments. These proceeded smoothly until he came to spermaceti, whereupon he performed a long series of experiments on this one substance.[56] If the minutes of the experiments correspond to the actual sequence, they suggest that spermaceti was also the starting point of Cavendish's complete understanding of a second rule of nature, that of latent heats. In the first of the spermaceti experiments, he mixed cold lumps with hot water, in the next hot melted spermaceti with cold water. From the measurements, he concluded that when spermaceti hardens it gives off heat, and when it melts it produces cold, and that the quantities of heat and cold are the same. He performed a series of experiments on another soft body, beeswax, which upon cooling changed from a hard cake to a thick syrup. He did further experiments of this kind on melted metals, bismuth, lead, and tin, cooling and solidifying them,[57] and on the change of state of ice to water, of water to steam, and back.[58] His varied experiments confirmed the following rule of latent heats: "As far as I can perceive it seems a constant rule in nature that all bodies in changing from a solid state to a fluid state or from a non-elastic state to the state of an elastic fluid generate cold, and by the contrary change they generate heat." Again he gave a physical explanation: "The reason of this phenomenon seems to be that it requires a greater quantity of heat to make bodies shew the same heat by the thermometer when in a fluid than in a solid state, and when in an elastic state than in a non-elastic state."[59] This explanation

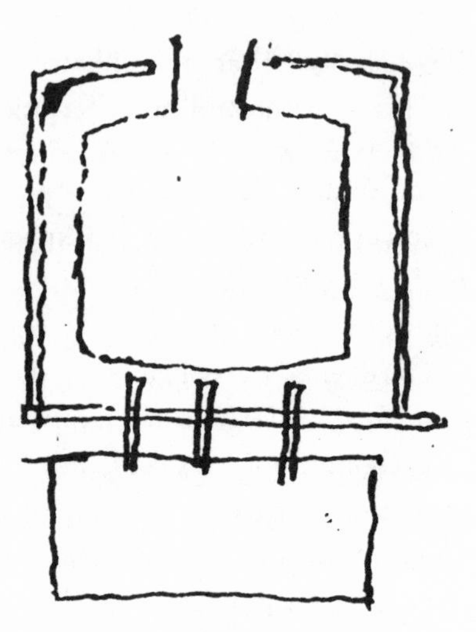

Illustration 11. Apparatus for experiments on latent heat. Drawing by Cavendish. In this experiment, the heat required to convert water to steam is measured by the time of evaporation of boiling water. Shown is a spirit lamp and a tin bottle surrounded by layers of brown paper insulation. Cavendish Mss III(a), 9:42. Reproduced by permission of the Chatsworth Settlement Trustees.

of the rule of latent heats was similar to that of William Irvine,[60] though Cavendish was critical of Irvine's theory.[61] With the two rules and their explanations, Cavendish now had a theory of both specific and latent heats.

In his investigation into heat, Cavendish adhered to the methods of the experimental philosophy. He began with the method of analysis: From experiments on particular substances such as water and mercury, he argued inductively for rules of specific and latent heats that apply universally to all substances. He then applied the method of synthesis: Having been guided to the rule of specific heats by experiments on liquids, he carried out new experiments on the other two states of matter, solid and gaseous; likewise, having been guided to the rule of latent heats by experiments on spermaceti and water, he carried out new experiments on, for example, the cold generated by the release of fixed air, or carbon dioxide, upon mixing alkalies and acids.

We speak of Cavendish's "theory," and although he labeled the folder in which he kept the paper discussed above "Experiments on Heat," our usage is correct for the time. As Newton had a theory of light and colors, Cavendish had one of specific and latent heats. Arrived at similarly by the methods of the experimental philosophy, Newton's and Cavendish's theories were comparable in their sweep: Newton's subordinated a great variety of observations of light, and Cavendish's brought together his own methodical observations with scattered observations on heat made by a number of experimenters over a long period. These included early observations by Fahrenheit and Taylor on the temperature of mixtures of hot and cold water and mercury, Cullen's observations on the cold generated by the evaporation of various liquids, French observations on the heat generated by freezing water, Black's observations on the heat generated by the condensation of steam in a distiller, and the familiar fact that when water begins to boil, it stays at the same temperature until it boils away.

From the first, in his investigation of the rules of heat, Cavendish kept before him the larger theoretical question of how to relate the two theories of heat, the theory of specific and latent heats and the mechanical theory of heat. In the minutes of his experiments from the 1760s, he juxtaposed the two: The change in temperature generated by a mixture of water and alcohol, he said, is caused either by the "commotion made by the particles of one uniting with those of the other" or by the "mixture of spt☉s & water requiring a greater quantity of heat to make it raise the thermom, to a given degree than the 2 liquors separately do." The first explanation referred to the mechanical theory of heat, the second to the theory of specific and latent heats.[62]

Hypotheses are absent from Cavendish's theory of specific and latent heats, but in an early draft, he stated the law of latent heats not as a "rule" but as an "hypothesis."[63] Between versions he changed his mind, possibly because with repeated experiments his confidence grew. "Rule in nature" conveys greater assurance than "hypothesis"; the former has proof, the latter wants it. He may have changed his mind also because of the form of argument he was using. By Newton's example, strictly speaking, a hypothesis does not belong in an inductive argument, as Cavendish's readers well knew.

Later Cavendish did introduce a hypothesis. In the 1770s, the decade between his two main theories of heat, he developed another, partial theory of heat. The occasion was a course of experiments on the adjustment of the boiling point of water on thermometers, carried out for the Royal Society. His principal colleague in the experiments was J. A. Deluc, with whom he had a sharp theoretical disagreement. Deluc identified heat with a fluid, the particles of which unite with the particles of water, and the two fly off together as steam, an explanation that Cavendish obviously could not accept. Writing up his own "theory of boiling," he gave it to Deluc for comment.[64] This paper, an outgrowth of his earlier paper on specific and latent heats, parts of which he inserted verbatim, proceeded from a numbered set of "principles" about the appearance of latent heat in evaporation and boiling, and the difference between the heats at which water boils when it is in contact with steam or air and when it is not, the "heat of ebullition." He first confirmed his principles experimentally; then with them he "explained" the main phenomena of boiling. His reasoning and presentation followed the usual methods of the experimental philosophy until the end, whereupon he gave an explanation of a different kind, a "hypothesis" about the "cause" of the difference between boiling and ebullition. The cause, a force that acts between the particles of water, alternates between repulsion and attraction, and the separation of the particles at which the alternations take place depends on the heat of the water. Elsewhere Cavendish attributed forces of this kind to Boscovich and Michell, and he discussed them in his mechanical theory, "Heat."[65] His theory of boiling combined in a single explanation two distinct

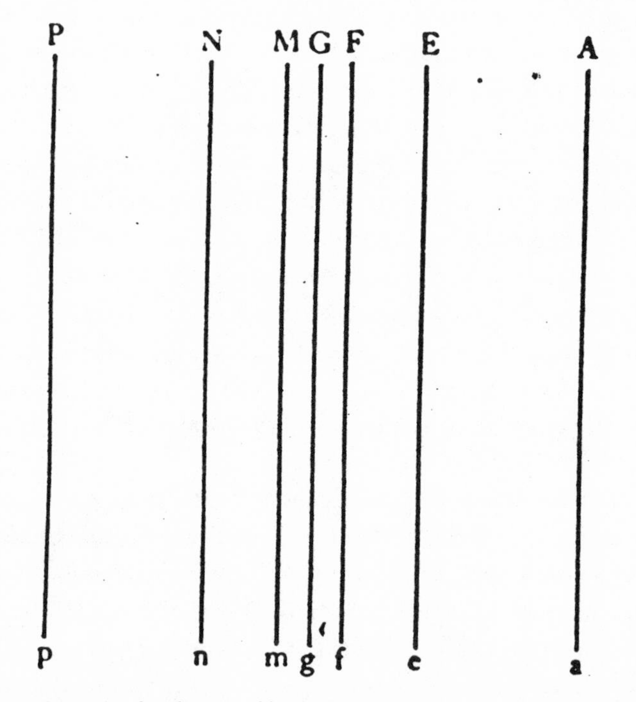

Illustration 12. Force in the theory of boiling. Drawing by Cavendish. PA represents a section of a flat plate of water. A particle is repelled by any particle closer than GM, and it is attracted by any particle between GM and GN. With this "hypothesis," Cavendish explained the "cause of the difference of the heat of boiling and ebullition." "Theory of Boiling," Cavendish Mss III(a), 5. Reproduced by permission of the Chatsworth Settlement Trustees.

kinds of theory, which he otherwise pursued separately. We can think of it as the first bridge between the two theories.

Publication

Cavendish did not publish his experiments on heat. The reason, it has been suggested, was that he did not want to enter into rivalry with Black.[66] That may be correct, but if so it is unlikely to be the whole explanation. There is rarely a worthwhile work in science that does not bring its author into rivalry of one sort or another, and in his first publication, on factitious air, Cavendish was not deterred by Black's prior work on fixed air; Black had even staked out a claim on the subject by saying that he intended to do more work. The one relevant difference is that Black published his original experiments on fixed air but not those on heat.

Another possible reason for deferring publication was that his experiments raised difficult problems for a mechanical theory of heat. He tried to resolve the problems, at first without success, and by the time he succeeded, there was no point in publishing. By 1780, a number of researchers were working with concepts of heat originating with Black, and before that, in 1772, Johan Carl Wilcke published his independent discovery of latent heat.[67] By 1783, if not earlier, Cavendish knew of Wilcke's as well as of Black's work, and in a publication that year, he invoked the rule of latent heat in a discussion of the freezing of water, giving neither an argument nor a citation for it, but simply remarking that it was a "circumstance now pretty well known to philosophers."[68]

Cavendish likely would have shown the draft of his experiments on heat to one or more colleagues before he laid it aside. It definitely was not lost to science, for through the experiments on which it was based, he acquired a thorough familiarity with heat, which served him well in his later researches. His first publication on heat was a paper on the freezing of mercury, by which time he had been studying heat for twenty years. Experiments on mercury had been carried out at the request of the Royal Society by a servant of the Hudson Bay Company, who was selected because of his favorable location, the frozen north. Thomas Hutchins's paper appeared in the *Philosophical Transactions* for 1783, followed by Cavendish's "observations" on Hutchins's experiments. These confirmed Cavendish's hypothesis that the great sinking of mercury in thermometers in extreme cold was due to the contraction of mercury upon turning solid. Cavendish presented his investigation as a direct continuation of his earlier unpublished experiments on freezing lead and tin and on the latent heat of water.[69] Not yet finished with the subject, he planned a second set of experiments to determine the greatest cold that can be produced by a freezing mixture of snow and various chemical solutions. The Hudson Bay experimenter this time, John McNab, produced cold "greatly superior" to any yet, as well as insight into the "remarkable" way nitrous and vitriolic acids freeze. Publishing McNab's experiments in 1786,[70] Cavendish asked him to do another set of experiments on the freezing of acids of varying strengths, which became the subject of his last published paper on heat, in 1788.[71] Cavendish intended to follow up these experimental studies with one more publication on heat, this one laying out his theoretical understanding of the whole subject. To this we now turn, the paper "Heat."

The Mechanical Theory Of Heat

Early Theoretical Statements

In his book on natural philosophy in 1782, William Nicholson discussed the nature of heat in the section on chemistry, a science, he said, which was still in its analytic stage, almost "purely experimental," consisting of "many facts and little theory."[1] Cavendish addressed this serious deficiency in natural philosophy with a theory of heat based jointly on his researches into specific and latent heats and on his researches into mechanics. Having examined these two lines of research separately, we now look at the theory that combined them. To bring together, as Cavendish did, two well-developed, general subjects, each with its own methods, concepts, and language, is one of the most difficult tasks of theoretical physics, and potentially one of the most rewarding.

When, in 1783, Cavendish discussed the freezing point of mercury, he did not use the term "latent heat" because it

> relates to an hypothesis depending on the supposition, that the heat of bodies is owing to their containing more or less of a substance called the matter of heat; and as I think Sir Isaac Newton's opinion, that heat consists in the internal motion of the particles of bodies, much the most probable, I chose to use the expression, heat is generated.[2]

He rejected Black's "latent heat" in this, his first public reference to the motion theory of heat. He did not give his reasons for saying that it was "much the most probable," not here or elsewhere in print. He did make one more public statement on the theory of heat, again to object to "latent heat," this in a paper on air and water the following year, where he remarked on a recent paper by James Watt on the same subject. In the passage in question, now remembered not so much for its bearing on heat as for its relevance to a priority dispute in chemistry, Cavendish gave his reasons for avoiding Watt's "language," Watt's "form of speaking":

> Now I have chosen to avoid this form of speaking, both because I think it more likely that there is no such thing as elementary heat, and because saying so in this instance, without using similar expressions in speaking of other chemical unions, would be improper, and would lead to false ideas; and it may even admit of doubt, whether the doing it in general would not cause more trouble and perplexity than it is worth.[3]

Cavendish's quarrel with the language of heat was joined in a general way by another theorist of heat: As in mathematical reasoning, James Hutton said, in scientific reasoning, terms must be clearly defined, and in theories of heat, scientific language was vaguest, a reason for the inability of investigators to agree.[4]

In his lifetime, the footnote on Black in 1783 and the passage on Watt the next year were all that Cavendish was to tell his readers about the nature of heat. Apart from "Heat," his scientific manuscripts contain two more references to Newton's theory of heat. One is discussed above, buried in a corollary to a theorem in a paper on the theory of motion, "Remarks," which reads, "Heat

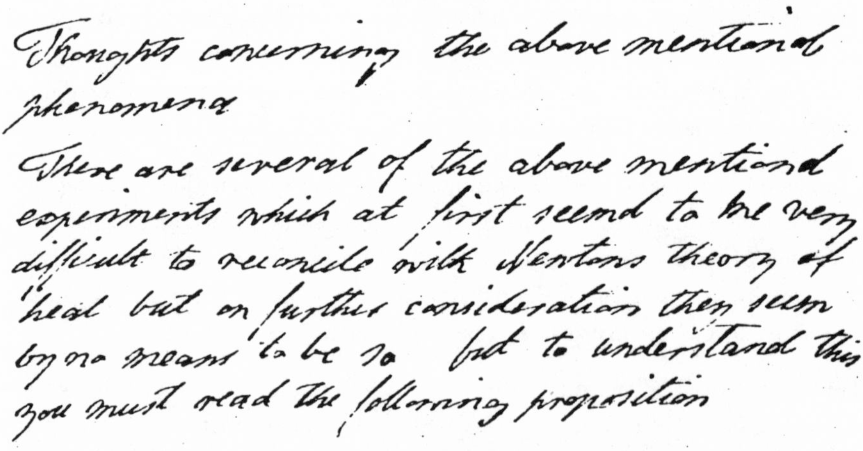

Thoughts concerning the above mentiond phenomena

There are several of the above mentiond experiments which at first seemd to me very difficult to reconcile with Newtons theory of heat but on further consideration they seem by no means to be so but to understand this you must read the following proposition

Illustration 13. Theory and experiment. With this passage, Cavendish concluded his paper on specific and latent heats. His mechanical theory of heat was, I think, the eventual follow-up. Cavendish Mss Misc. Reproduced by permission of the Chatsworth Settlement Trustees.

most likely is the vibrating of the particles of which bodies are composed." The other reference appears in his paper on specific and latent heats, which concludes with the observation that certain of his experiments at first seemed to him "very difficult to reconcile with Newton's theory of heat, but on further consideration they seem by no means to be so. But to understand this you must read the following proposition."[5] Unfortunately, there the paper ends, abruptly, without the promised proposition. Until recently these references, two published and two unpublished, were the only known explicit statements by Cavendish on the theory of heat. Because it can be shown that Cavendish's understanding of the nature of heat entered into his researches on factitious airs, the production of water, and electricity, as well as his researches on the freezing of mercury and on freezing mixtures,[6] what was missing was a fully developed theory of heat, one comparable to his fully developed theory of electricity.

Preparation

Try as they might, Black and his followers could form no idea of the internal motions of bodies capable of accounting for the phenomena of heat. Even in the case of friction, which offered the strongest support for the mechanical theory, they found it hard to picture the motions responsible. Most telling, in Black's view, was the failure of supporters of the mechanical theory to show that motion could explain the entirety of the phenomena of heat. The same complaint could not have been made about the material theory, certainly not after the dissertation by Black's student, Cleghorn. Cavendish set about to supply what was missing from the side of the mechanical theory.

Cavendish's belief that mechanics was the eventual route to a proper understanding of heat was of itself unremarkable. What set Cavendish apart from his colleagues was that he acted upon his belief; that, despite the many compelling arguments against the mechanical theory, he developed it with the object of publishing: and that, given the stage of natural philosophy, he was able to carry the theory as far as he did.

He could take heart from his earlier theory of electricity. The phenomena of electricity and the phenomena of heat—shock, attraction, and polarity on the one hand, and warmth, expansion, and change of state on the other—were not at all similar, but their understanding was. From the 1760s, both subjects made use of a pair of fundamental magnitudes representing a quantity and an intensity: charge and potential, and heat and temperature. The quantity in each subject had a latent and an active state and obeyed a conservation law, and each subject assigned specific, permanent properties to different substances. Given the parallels between electricity and heat, and given the common natural philosophy from which they sprang, we might expect that if one could be explained by a mechanical analogy, so could the other. Cavendish showed that this was indeed the case: having developed a mathematico-mechanical theory for electricity, he intended to do the same for heat. We might further expect him to have continued with the same analogy, to have made a mathematical theory of the fluid of heat as he had of the fluid of electricity, but we know differently. Having rejected the fluid theory of heat, he needed a new approach. With the theory that resulted, as far as we know his last, Cavendish brought the mechanical understanding of heat to a level that would not be surpassed for more than half a century.

To assemble a mechanical theory of heat, Cavendish needed to make decisions; natural philosophy supplied him with his options. To begin with, he had to decide on the scope of his project. He could set out to explain heat together with the other powers of nature, within a single scheme. In natural philosophy in the middle of the eighteenth century, comprehensive systems had been common, an approach that conceivably could have interested Cavendish, even in the 1780s, but it did not. Long before, to his way of thinking, Maclaurin had got it right when in his book about Newton's discoveries, he said that natural philosophy no longer took in the "whole scheme of nature" at once, encompassing everything in a single view, but studied nature "in parts," proceeding by "just steps."[7] Cavendish restricted his theory to one part of nature, heat, and referred to other parts only as they bore on the subject of interest.

Within that part, Cavendish set out to make a comprehensive theory, and that meant he had to decide what constituted the field of heat. To a considerable degree, the decision had been made for him; by the 1780s, despite their different emphases, natural philosophers and chemists largely agreed about the principal phenomena of heat.

Early on in the plan, Cavendish had to decide whether to present the theory by itself or to present it together with a report of experiments. Related to this decision was another, whether to treat the theory at book length or succinctly as a paper. Cavendish had carried out extensive experiments, which he originally intended to present together with a proposition on "Newton's theory of heat," perhaps followed by his mechanical theory. In "Heat," he chose not to include them; heat having been a major experimental field of natural philosophy for twenty-five years by then, he could count on his readers to be familiar with the experimental basis of the theory. The publication of a purely theoretical argument was uncommon, but not unprecedented. Cavendish needed no more than the space of a paper to develop his mechanical theory of heat.

With regard to the kind of theory, Cavendish had to make two basic decisions. First, he could develop the theory either mathematically or qualitatively; second, he could develop it either from a causal hypothesis or from a law. Earlier Newtonians had viewed mathematical demonstrations and hypotheses about causes as antithetical ways of going about their work, favoring the former as a corrective to the latter, which they identified with "conjectural Philosophy," not "natural Philosophy."[8] Cavendish's choice, a mathematical and causal theory, reflected a change in thinking in natural philosophy.

In deciding on the mathematics to use, Cavendish followed precedent. The standard British practice was to represent mechanical problems by geometrical figures and, if the calculus was needed, to take infinitesimal figures to the limit, using Newtonian fluxions as a convenient calculational technique.[9] In "Heat," Cavendish derived the law of conservation of vis viva using infinitesimal geometry and fluxions. His reasoning, straightforward then, takes careful reading today. We are inclined to disagree with an author of mathematical papers on gravitational astronomy in the *Philosophical Transactions* who judged geometrical reasoning, Newton's way, as much more elegant, rational, and simple than the analytical calculus.[10] Today we prefer Leibniz's way to Newton's.

As in mathematics, in mechanics Cavendish followed British practice, which again looks rather strange to us. Had he been European, he might have used a more familiar form of mechanics. In or about the same year that Cavendish wrote "Heat," J. L. Lagrange published *Mechanique Analitique*, in which not a single geometric diagram appears, the first important work in mechanics to make this claim. Lagrange carried out every proof in the analytical manner of the Leibnizian calculus, an implicit assertion of the superiority of algebraic over geometrical methods. He deduced the theory of mechanics from the statical principle of virtual work and from d'Alembert's dynamical principle, and he expressed the differential equations of motion, known today by his name, in quantities later called "kinetic" and "potential energy."[11] Although the principle of conservation of kinetic and potential energy appears in both his and Cavendish's mechanics, it is a limited agreement. The mechanics of "Heat" is a mechanics of energy, but it is not the mechanics of the European mathematicians. It is Newton's mechanics enlarged to include vis viva as a mechanical effect of a body in motion.

Whatever Cavendish decided upon as the cause of heat, with it he had to explain why heat is apparently conserved, why when two bodies at different temperatures are combined, one body gives up as much heat as the other gains. The material hypothesis trivially explains it, since a defining property of the matter of fire is that it cannot be increased or destroyed. To explain it by motion, mechanics offers two conserved quantities, momentum and vis viva, the former a quantity made up of parts, a magnitude and a direction, the latter a simple, absolute magnitude. The difference between the two bears on the following objection to the mechanical theory of heat, as it was discussed by Isaac Milner. The quantity that is conserved in simple heat exchanges has also to appear and disappear in changes of state, chemical changes, and combustion. Momentum cannot be created from nothing, for otherwise a perpetual motion machine would be possible; therefore, the argument went, heat cannot be motion. As a defender of the mechanical theory, Milner got around this objection by identifying heat with only part of the momentum, the magnitude, leaving out of consideration the direction. This, the "absolute Motion" of the particles, does not obey the law of conservation of momentum; for example, the total directed momentum of two equal particles moving in opposite directions with equal speed starting from a point of rest remains zero, but the particles acquire an absolute motion equal to the sum of their absolute momenta. Milner ignored the promising alternative, the second conserved measure of motion, vis viva, which can make a sudden appearance, arising not from nothing but from potential or latent vis viva. Milner discussed the above objection in his lectures in Cambridge in the mid 1780s.[12]

To make a theory of heat, Cavendish had also to make decisions about the general concepts of natural philosophy, force, motion, and matter. For the accelerating forces responsible for heat, predictably he chose attractive and repulsive centripetal forces. For the kind of motion to identify with heat, he chose the likely candidate, the internal motion that retains the form of bodies, vibration. Vibration, was thought to be the motion favored universally by nature, as pointed out by Robert Smith, Professor of Astronomy and Experimental Philosophy in Cambridge: "[A]lmost all sorts of substances are perpetually subject to very minute vibrating motions, and all our senses and faculties seem chiefly to depend upon such motions excited in the proper organs."[13] In his hypothesis of the cause of heat, Cavendish spoke of "internal motion," but he understood that the motion in question is vibration.

There were two parts to Cavendish's decision about matter. The first dealt with matter in general: On this disputed subject, Cavendish thought that Boscovich was undoubtedly right, that matter consists of central points surrounded by alternating spheres of attraction and repulsion. That was the extent of his agreement with Boscovich, who otherwise did not have a conservation law, regarded vis viva as having little significance, and believed in the matter of fire.[14] The second part of the decision was about the particular kind of vibrating matter. To the question of what it is that vibrates, a variety of answers had been proposed: The particles of the air and acid sulfur in bodies, those of an universal aether, those of fire, and those of ordinary bodies. We consider the last three. In the General Scholium of the *Principia*, Newton speculated on an "electric and elastic spirit" pervading all bodies, the action of which causes all attractions and repulsions, specifically those responsible for heat, and in *Opticks* he speculated on an "Aether."[15] Cavendish never used the word "aether," and it would seem that early in his career he dismissed the possibility of an electric, elastic medium: In a discarded version of his electrical theory, he developed the mathematical consequences of an electric fluid filling infinite space, concluding with a discouraging note, "how far that supposition will agree with experiment I am in doubt," and he may have had mathematical doubts about such a fluid as well.[16] In any event, he did not introduce an aether into his theory of heat. Nor, having previously denied its existence, did he introduce a specific fluid of fire or heat. Instead, he identified the vibrating matter with the discrete particles of ordinary matter.

The Mechanical Theory

Cavendish's eighteenth-century universe was well ordered, balanced, permanent, and stable. Conceived of as matter moving according to immutable laws, it had no built-in direction. Although change contributed to the panorama of nature, it was a superficial feature; as every change was compensated by a contrary change.[17] This presupposition of thought entered natural philosophy, and it was within its possibilities and limitations that Cavendish formulated his theory of heat. We see it in the facts he addressed: equilibrium, reversibility, equivalence, and conservation. We see it in his selection of concepts, which together ensured the balance of nature and the universality of the laws of nature or motion: attraction and repulsion, visible and invisible vis viva, actual and potential vis viva, and active and latent heat.

For the force of moving bodies, Cavendish used the term "vis viva," which was standard. In his early study of vis viva, "Remarks," instead of that term, he used his own, "mechanical momentum," picking up on the parallel between Leibniz's and Newton's forces of moving bodies. In "Heat," the parallel was implicit. Newton proved that the momentum of the center of gravity of a system of bodies is unchanged by the "actions of the bodies among themselves";[18] Cavendish proved that the vis viva of a system of bodies is unchanged by "their motions among each other." That, as it happens, was the appropriate formulation of the law of conservation of energy for the theory of heat as motion.

"Heat" begins with theorems on the additivity of the vis viva of the several possible motions of a body. The extent to which Cavendish elaborates on this property is an indication of the unfamiliarity of the theory of vis viva among his prospective British readers, as it is an indication of its importance, which derives from the undirected, or scalar, character of vis viva. He divides vis viva into two kinds, "visible" and "invisible." The visible is the vis viva of the center of gravity of a body undergoing progressive motion or of the body undergoing rotation, or both; the invisible is the vis viva of the particles of the body moving among themselves; and the total vis viva of the body is the sum of both. The invisible vis viva is further divided into two parts, actual and potential (not Cavendish's term). The symbol s stands for the actual vis viva of all of the particles constituting the body; the symbol S stands for one-half the sum of the vis viva that each particle would acquire by the attraction or repulsion of every other particle in falling from infinity to its actual position within the body.

Following upon the above considerations, Cavendish derives the law of conservation of vis viva or, in our terms, the law of conservation of energy, kinetic and potential. In his earlier study of vis viva, "Remarks," he derived the law with help of a theorem of the *Principia*, which states that if two bodies are moved by the same central force, and if they have the same velocity at any one given distance from the center, they will have the same velocity at all other equal distances, regardless of the paths they take to get there.[19] In "Heat," he assumes the same kind of forces as in "Remarks," although this time he does not explicitly invoke Newton's theorem. The law he derives states that the total vis viva of the particles of an insulated body, the sum of actual and potential vis viva, the quantity s − S, cannot change owing to motions of the particles among themselves. Because actual vis viva arises from the motions of particles, and potential vis viva from their separations in space, what is conserved is not the motion of the particles, but a more complicated, abstract function of their mechanical state.

Daniel Bernoulli arrived at a similar result. Both he and Cavendish analyzed simple cases and then generalized to any number of mutually attracting bodies, and both employed a concept of potential vis viva and admitted attractions obeying indeterminate laws, but their mathematics, notation, and terminology were different. Bernoulli expressed his final result as the composition of the vis viva of the moving bodies from what he called the "partial *vis vivae*" of all the interactions of the bodies taken two by two.[20] If Cavendish knew of Bernoulli's derivation, his own was not guided by it. As in "Remarks," in "Heat" Cavendish does not mention Bernoulli's name.

The second part of "Heat" introduces the subject proper, the theory of heat. Cavendish makes the transition by means of a hypothesis and a pairing of terms, allowing him to compare vis viva with heat. The "Hypothesis" reads, "Heat consists in the internal motion of the particles of which bodies are composed." The "active heat" of a body is the actual vis viva, s, of its particles; the "latent heat" is the potential vis viva, − S; and the "total heat" is s − S. The total heat of a body is conserved because the total vis viva is. Not identified with a specific mechanical quantity, the "sensible heat" of a body is the heat according to the thermometer. The "capacity" of a body is the total heat required to raise the sensible heat of a given weight of that body by a given amount. In "Heat," Cavendish accepts what he previously had not, Black's way of talking about heat, "latent heat" and "capacity."

Changes in the total heat in a body, s − S, can be measured, but not the active and latent heats, s and − S, individually. For instance, in the communication of heat from a hot to a cold body, how the heat is divided between s and −S in the two bodies depends on "some function, either of the size of their particles or of any other quality in them" such as the separation of the particles or the frequency of their vibration. In the state of science, this detailed knowledge of the interior of bodies is clearly inaccessible.[21] Nevertheless, with the help of s and S, the mechanical theory can explain the principal phenomena of heat very well. This is what Cavendish sets out to demonstrate in the remainder of "Heat."

The Question

Experiments

As he did heat, Cavendish interpreted light in Newton's way: light is a body consisting of particles emitted from luminous bodies at a known velocity. To find its vis viva, he looked to Michell's experiment with a light-mill.[1] From the observed speed of rotation of the vane and other details of the experiment, and from the assumption that light is perfectly reflected from the copper of the vane, he calculated the momentum and vis viva of sunlight falling each second on one and a half square feet of surface. Translating this result into its mechanical effect, he showed that the rate of vis viva of sunlight falling on that surface exceeds the work done by two horses, two horsepower.[2] By his theory, this very considerable mechanical power imparted by sunlight has an exact heat equivalence: if the same quantity of light were to be absorbed instead of reflected, the equivalent heat would register on the thermometer. Upon this reasoning, Cavendish proposed two experiments on sunlight, with the intention of carrying them out himself.

The first was to "expose thermometers whose bulbs are coated with various dark & equally dark colourd substances alternately to the ⊙ [sun] & shade & see whether they receive the same increase of heat in the same time." This experiment would test and follow up what he called "a necessary consequence of this theory." Cavendish gave no details, but the general kind of experiment he had in mind had recent predecedents. The Cambridge Professor of Chemistry exposed a thermometer to the sun, and then after blackening its bulb, he exposed it again, noting a higher reading the second time. He hoped that others would paint their thermometers with different colors to determine the disposition of colors to receive and retain heat.[3] Acting on this suggestion, the Royal

Society's Bakerian Lecturer exposed a thermometer painted with different colors to determine if the absorption of heat follows the progression of the prismatic colors, or if it follows some other law. He also painted a thermometer black, and after placing it alternately in the sun and the shade, concluded that every degree of light is accompanied by a proportionate degree of heat.[4] In 1787, about the time that Cavendish wrote his paper, George Fordyce reported on experiments to see if sunlight falling on blackened bodies of different substances heats them equally. The results were indecisive, but from this experiment and other facts, Fordyce concluded that heat could not be material, that it must be a "quality," which might or might not be vibrations.[5]

The second experiment was to measure the heat equivalent of the vis viva of sunlight. Foreshadowed in the rough sketch of the paper, "Calculation of vis viva of $\odot^{s}$ rays & D^{o} required to commun. given quant. heat," the experiment is described in the revised copy, "Exper. to determine the vis viva necessary to give a given increase of sensible heat to a given body by alternately exposing a thermometer in the ⊙ & shading."[6] Given the physics of the next century, this is the experiment we wish Cavendish had carried out, as perhaps he did. In any case, this experiment, combined with Michell's light-mill, would have yielded a numerical value for the mechanical equivalent of heat. Because the light-mill was misinterpreted, the value would have been wrong, but that is beside the point here. Cavendish's object was to determine a physical constant, which would have joined a small number of other useful constants such as the velocity of light in optics, and the acceleration of free fall in mechanics. He already had a use in mind, to estimate the "velocity with which the particles of a body vibrate."[7] The constant he spoke of, the equivalence, which we know to be universal, was correctly and accurately determined sixty years later.

Although Cavendish uses similar words, "mechanical equivalent of heat" is our term, not his. We should note the anachronism, lest we be misled. Nowhere does Cavendish's wording suggest that he had in mind a universal constant. By the same token, had the idea presented itself to him, we have no reason to think he would have rejected it. The point bears examining.

A determination of the mechanical equivalent of heat would have strengthened Cavendish's theory of heat. With it, he would have made the hypothesis of his theory of heat quantitatively complete, as he had done with his hollow-globe experiment in making the hypothesis of his theory of electricity complete. He would also have addressed a well-known problem of natural philosophy, violations of the principle of conservation of actual and potential vis viva. Lost vis viva had been attributed to invisible motions or internal deformations or some other hidden process, but no one had suggested heat, and no one produced experimentally a quantitative measure of the transformed loss.[8] Cavendish intended to do just that, to produce a quantity of heat as the measure of the transformed vis viva.

The quantitative concepts necessary for expressing the mechanical equivalent of heat were available. Cavendish's measure of vis viva was mechanical work, the lifting of a weight through a height, our foot-pound, and his measure of quantity of heat was the heat required to raise the temperature of a given weight of water one degree by the thermometer, our BTU. Papers proving otherwise might come to light one day, but let us assume that Cavendish did not make the determination. We may ask why. The reason might well be Michell's experiment, upon which Cavendish's experiment depended for the vis viva of light. With or without Bennet's repetition of the experiment, Cavendish may have realized that Michell's light-mill was not powered by the momentum of the light striking it, that a more likely explanation lay in the subject to which he was then giving full attention, heat. Michell, after all, had actually melted the copper vane with the concentrated rays of the sun, disabling the apparatus. Today's explanation of Michell's light-mill requires highly sophisticated physics, unavailable in the eighteenth century. Although the modern electromagnetic theory of light requires that light exert a pressure on the vane, it is minute compared with convection and radiometer effects.

Alternatively, if Cavendish had come to doubt Michell's experiment, he could have turned to a mechanical source of heat, a direct route to the mechanical equivalent of heat. Later James Prescott Joule would take that route in what he regarded as the simplest and most persuasive proof of the existence of this constant of nature: With a paddle wheel in water driven by descending weights, a precision thermometer, and an accurate technique for measuring small differences of temperature, he correlated work and heat.[9] If Cavendish had considered experiments of this kind, he might have rejected them, judging the heats involved as analogous to the heats produced by the emission of light, too small to measure. Thomas Young, although a supporter of the motion theory of heat, observed that fluids were incapable of acquiring any material increase in temperature by internal friction,[10] a circumstance cited as an argument against the motion theory of heat.[11] Cavendish did propose an experiment on the mechanical production of heat from friction, "exper. whether friction is as much diminished by oil & grease as the heat is," but not obviously with the intention of measuring the mechanical equivalent of heat.

There is a theoretical reason, too, why Cavendish might not have measured the mechanical equivalent of heat. Because he did not express theoretical results as equations between terms with physical units, conversion or equivalence factors did not come up as a matter of course. He did not determine the gravitational constant *G*, either, although he readily could have, and in the next century his experiment of weighing the world was repeated with that end in view. If we keep in mind that universal constants did not have the place in science then that they have today, Cavendish's evident failure to pursue the mechanical equivalent of heat is not all that puzzling.

Our conclusion is reinforced if we compare Cavendish with Benjamin Thompson, Count Rumford, whom we have met as a former believer whose faith in the fluid hypothesis of heat was shaken by experiments on the weight of heat. In the *Philosophical Transactions* in 1798, Rumford reported his celebrated experiment on the production of heat by friction. Pressing a dull, steel borer against the metal cylinder of a cannon barrel turning at a given rate, in less than three hours he raised the temperature of nearly twenty pounds of water from sixty degrees to the boiling point. This considerable heat did not arise from a change in the specific heat of the metal or from the air, and it seemed inexhaustible. Rumford concluded that heat could not be a fluid, that it must be motion, the internal vibrations of bodies. He reported how much heat could be produced by a mechanical contrivance, and he even made a rough estimate of the mechanical equivalent of heat, but he emphasized the negative implication of his experiment, its disproof of the fluid of heat, and that was how his paper was read.[12]

Cavendish studied heat at a time when importance was placed on a decision between the mechanical and fluid hypotheses, even if researchers could and did work with laws and measurements independently of them. Around the time he wrote his theory, a colleague said that the nature of heat was currently a "great question" of natural philosophy.[13]

Rumford's experiment points to another possible reason why Cavendish did not have a wide-ranging plan to determine the mechanical equivalent of heat. This has to do with the content of his theory. Although his theory requires the existence of an equivalent, the heat that is produced or lost in an exchange of vis viva is a quantity of total heat, part of which is latent heat, which does not register on the thermometer. Consequently, different sorts of experiments would be expected to give different values for the constant. Cavendish makes the point in his discussion of heat generated by friction between solid bodies: the loss of vis viva of the bodies in frictional contact is matched by an "augmentation of total heat equivalent thereto," but this loss is attended by a displacement of particles and with it an alteration of latent heat, "which will commonly make the alteration of sensible [thermometer] heat very different from what it would otherwise be." By this reasoning, the numbers recorded in Rumford's cannon-boring friction experiment might not yield a reliable value for the mechanical equivalent of heat. By contrast, Cavendish's planned experiment on the absorption of sunlight would minimize any complications arising from changes in latent heat. Experiments suited for determining the mechanical equivalent of heat would have to be chosen carefully.

We close with a consideration of Cavendish's planned experiments in relation to the first law of thermodynamics. That law is the conservation of energy, the great unifying law of physics. In establishing it, a first step is to determine the quantitative relationship between energy, or work, and heat. It is necessary to show that all of the experimentally accessible ways of transforming work into heat obey the same equivalence, and that the transformation is independent of the agency effecting the transformation. That is what Cavendish's two proposed experiments were about. It is also necessary to show that in the reverse transformation, that of heat into work, an equivalent quantity of heat always disappears; Cavendish did not propose an experiment with this second object. With his single instance of a one-way transformation, he had only started on the problem of establishing the mechanical equivalent of heat. That, together with the paucity of quantitative laws of nature then available to him, meant that he could not have treated the problem thoroughly. When, sixty years later, the mechanical conservation law was again examined in relation to heat, the equivalence relation was properly addressed, and only then did the theoretical power of the conservation of energy become evident.

Hypotheses of Heat

Where Rumford looked to experiment, Cavendish looked to theory to answer the question of the nature of heat. Passing from one branch of natural philosophy to another, he showed that Newton's view of heat did "really explain" the phenomena of heat. "Heat" was a continuous argument for the hypothesis that heat consists of the internal vibrations of bodies.

By rigorous and, where possible, mathematical arguments, Cavendish proceeded from a mechanical hypothesis together with received mechanical principles to demonstrate a detailed analogy between the effects of vibrating particles and the phenomena of heat. By showing that the hypothesis is fully sufficient to explain the phenomena, he established one half of the argument. To establish its necessity, the other half, he called implicitly upon the principle of causality. The vibrations of particles surely take place, he reasoned, and these vibrations must be a cause and, as such, must produce effects that present themselves to our senses as phenomena; the effects of vibrations are analogous to the phenomena of heat, and no other phenomena can reasonably be attributed to this cause.[14] With that, Cavendish let rest the case for Newton's view of heat. Confirmed by its consequences, and having independent empirical plausibility, the mechanical hypothesis met the test of a good hypothesis. Cavendish's theory corresponded to the best theoretical reasoning of the time.

Cavendish referred explicitly to the material hypotheses of heat only at the end of his paper, and even there he did not discuss it:

> [T]hough it does not seem impossible that a fluid might exist endued with such properties as to produce the effects of heat; yet any hypothesis of such kind must be of that unprecise nature, as not to admit of being reduced to strict reasoning, so as to suffer one to examine whether it will really explain the phenomena or whether it will not rather be attended with numberless inconsistencies & absurdities.

Three times in the conclusion, Cavendish used the expression "strict reasoning." Although in another subject, electricity, Cavendish demonstrated that a fluid hypothesis lent itself to, as he said, "strict reasoning," in heat it failed the "test."[15] In a theoretical study such as his, the critical test of a hypothesis was not its agreement or disagreement with the data, but the strictness of the theoretical reasoning.

According to a commentary at the time, defenders of the hypothesis of heat as a fluid had to show that it accounted for all of the phenomena, and also that the phenomena could not be explained without it.[16] By demonstrating that the opposing hypothesis could explain the facts, Cavendish removed the second of the two pillars on which the fluid hypothesis rested. The latter might meet the first criterion, but that was only because it was "pliable," "easily adapted to any appearances."[17] "Nothing proved the existence" of such a fluid, Cavendish concluded.

Had Cavendish been a scientific instrumentalist, concerned only that a theory predicts the outcomes of experiments, he would not have bothered to develop an alternative to a theory that worked, or if he had, he would have gone about comparing the theories differently, concerned to show that they did not work equally well. Rather, he believed that his hypothetical theory of heat was probably true of the external world, that Newton's invisible vibrations were a reality. He had arrived at the threshold of the molecular theory of heat.

Omissions

In "Heat," Cavendish considered the production of heat by a variety of agencies, but with the exceptions of the expansion of bodies and chemical decomposition, he ignored the effects of heat. He did not, for example, discuss the steam engine, which demonstrated the converse of what he proposed to measure: Instead of work into heat, it transformed heat into work. The heat engine acquired theoretical significance in the nineteenth century in connection with the second law of thermodynamics, but Cavendish did not foresee that law.

Nor did Cavendish discuss magnetism, which was known to be affected by heat; for example, the attractive power of a magnet decreases when it is immersed in boiling water,[18] and magnets differ among themselves according to differences in the heats to which their steel is subjected in hardening.[19] Unlike electricity, magnetism could not be collected, experimented with, and quantified independently of its source. Restricted to iron and substances containing iron, magnetism seemed to be a special, not a universal, property of matter, considerations that definitely hindered its theoretical understanding; of his experimental fields, it was only magnetism on which Cavendish did not express a theoretical opinion. To have incorporated magnetism into his theory of heat, he would have had to connect it with the motion of particles, for which he needed a theory. No doubt his main reason for omitting this important branch of natural philosophy was that magnetism does not generate heat in the way that moving electricity does.

Cavendish ignored the electrical effects of heat, too. He examined the heat generated by electric current, but not the electricity generated by heat, a phenomenon known since the middle of the century. When immersed in heated water, a specimen of the mineral tourmaline acquires a positive charge at one end and a negative charge at the other, and upon cooling, the opposite charges appear. Having observed that tourmaline conducts electricity in only one direction, an experimenter drew an "analogy" between it and a magnet, which passes magnetism in only one direction.[20] The analogy between tourmaline and magnetism, with its singular property, suggests a reason why Cavendish omitted pyroelectricity from his theory of heat.

"Heat" contains no discussion of air, which might seem surprising, since air was Cavendish's great experimental subject. Convection currents, the motion of air arising from differences in heat, had been considered by Euler, but Cavendish ignored the effect.[21] Nor did he consider, as Joule would, the heating effect of the compression of air, even though he had experimented extensively on this and related phenomena.[22] The explanation would seem to lie in his brief remarks on the expansion of bodies in general, which implied the effect; he must have seen no need to discuss it separately. It is instructive to compare him with Daniel Bernoulli on this point. By analyzing the increase in the number of collisions of the particles of a confined air when it is compressed, Bernoulli derived the inverse proportionality between the pressure and the volume of air, Boyle's law. Further, recognizing that not only compression but also heat increases the pressure of air, he showed that pressure varies with the square of the average velocity of the particles of air, or vis viva, but he stopped short of identifying this vis viva with temperature. If Cavendish had not read Bernoulli's theory, he probably would have read Deluc's summary of it in his treatise on meteorology,[23] but even so, we should not wonder that he failed to refer to it. Bernoulli's excursion into the theory of air was very brief, and although he considered an effect of heat, he did not develop a theory of heat. Bernoulli's analysis was based on translational motion and impacts; Cavendish's, on vibratory motion and continuously acting forces. His approach to heat did not point in Bernoulli's direction, the kinetic theory of gases.

To readers of Cavendish's paper today, its most serious omission is probably the absence of theoretical predictions. The paper makes retrodictions, and although they support the theory equally as well as confirmed predictions, to our way of thinking they lack the same psychological force. No doubt one reason he did not make explicit predictions was because he did not think of science in the first instance as being about predictions. Looking to science for understanding, he wrote "Heat" to decide between rival explanations of heat, to show that the Newtonian theory was probably true.

A maker of theories in Cavendish's time had less incentive than a theorist today to publish explicit predictions. Usually he was his own experimenter. For Cavendish to have made predictions about heat without following them up himself would have been to leave his work in an incomplete form. In his own good time, he would have carried out any experiments implied by, or otherwise bearing on the truth of, his hypothesis of heat, the subject for a follow-up publication.

When the theorist and the experimentalist were no longer the same person, the theorist's job came to be to interest the experimentalist. In this later stage of physics, theoretical papers might extract experimental implications and state them explicitly as predictions, as the point of entry for experimentalists. Cavendish lived in his own time, and his way of dealing with theory and experiment reflected that fact.

Why

For reasons spelled out in the appendix, it is probable that Cavendish wrote "Heat" in the late 1780s. As to the immediate stimulus for writing it, he said nothing. At the time, he was in touch with the engineer John Smeaton on the subject of momentum and vis viva, but this had nothing to do with heat.[24] More to the point, in 1783 he received a paper on calorimetry by P. S. Laplace and A. L. Lavoisier.[25] Its authors presented the motion and material hypotheses of heat side by side, without deciding between them, keeping only what both theories had in common, the principles of conservation of heat and of reversibility of heat in changes of state. Of the two, it was Laplace who formulated the motion hypothesis: "[H]eat is the vis viva resulting from the imperceptible motions of the constituent particles of a body." Which hypothesis Laplace actually preferred is uncertain, and he was later to hold the material theory of heat, as Lavoisier did.[26] What Laplace had to say about the motion hypothesis in 1783 was very brief and nonmathematical and contained no theoretical development, but in it Cavendish would have met a reflection of his own reasoning. He read Lavoisier and Laplace's paper on heat with critical interest.[27]

At about this time, George Fordyce and Adair Crawford announced to the Royal Society a series of experiments showing that heat diminishes the gravity of bodies; Fordyce followed with his paper on the loss of weight of ice upon melting,[28] and although his finding was credible, the experiments were delicate and subject to errors that were hard to control. Charles Blagden, Cavendish's close associate at the time, wrote to Laplace about Fordyce's experiments, and to C. L. Berthollet asking for information on recent, similar experiments in France, and he kept Cavendish informed on the subject.[29] Cavendish definitely was interested, having witnessed earlier experiments of that kind.[30] It seems unlikely, however, that Laplace and Lavoisier's paper or Fordyce's paper or any other theoretical or experimental paper on heat was the reason for Cavendish to write "Heat," or he would have mentioned it. Nor was "Heat" based on a new understanding of his. The central idea, the identification of heat with vis viva, had likely come to him long before.

We may never know for certain why Cavendish wrote a theory of heat, but a plausible, if partial, reason is that he had recently been doing experimental work on heat and now wanted to clarify for himself, anew, the foundations of the subject, and to confirm the adequacy of the theory overall. Nothing in his wording indicates that he was in serious doubt about the truth of his hypothesis, that heat is the vibration of particles, but he had questions. The sketch of the paper has a tentative feel, for example, "Heat by friction & hammering Whether they can give suffic vis viva."[31] He had an additional motive for writing his theory, of this we are certain. Not meant for his eyes alone but for publication, "Heat" was written to show the scientific world how to understand heat in Newton's way.

We learn more about Cavendish's motivation by looking into the state of natural philosophy at the time. Natural philosophers had a general interest in the kind of question he addressed, that of cause or power. Their interest in this was sharply focussed in three leading fields of research: in pneumatic chemistry, in optics, and in heat, and perhaps also in a fourth, electricity. Phlogiston as a causal principle or substance was challenged by the new antiphlogistic chemistry, based on oxygen. Phlogistic and oxygenic theories of combustion were incompatible; if one theory was right, the other was wrong. In optics the alternatives were not as clear-cut. Attempts had been made to combine the particulate and vibrational theories of light, and there were several fluid theories in addition. It was increasingly clear that no existing theory was fully satisfactory; yet one or the other main theoretical direction in optics, the particulate or the vibrational, was widely expected to correspond to reality.[32] Likewise, in heat, there were two main directions, the motion and the material theories, with variants. Natural philosophers might proceed with their experiments on heat while remaining undecided about its cause, but they usually had a preference, and they fully expected the issue to be clarified one way or the other.

The publication in Britain of several important books on heat around 1780 highlighted the question of the cause of heat, and the *Philosophical Transactions* published an uncommon number of papers on heat in 1787–88. Cavendish may have concluded that it was time to settle the issue, especially as the motion theory of heat had not been heard from, while support for the fluid theory was evidently growing. William Nicholson commented in his treatise on natural philosophy in 1782 that the view of heat as the vibration of particles was "scarcely hypothetical," that the hypothetical fluid of heat demanded "amazing," scarcely credible properties, and that to postulate a fluid of heat was to multiply causes in violation of the rules of scientific reasoning. Eight years later, in his treatise on chemistry, Nicholson left undecided the nature of heat, instead drawing attention to it as a problem deserving of the attention of natural philosophers.[33]

In 1786 the chemist Bryan Higgins, in a book on the latest advances in heat, light, and pneumatic chemistry, did not justify his use of the material view of heat on the grounds that Cavendish together with Black and other distinguished natural philosophers "have accepted it."[34] Higgins was almost certainly in error about Cavendish, but if Cavendish took notice, he would have realized how incompletely he had made known his contrary, beleaguered Newtonian view of heat.

At the time of "Heat," the understanding of physical reality that had guided Cavendish in his researches was everywhere under attack or ignored. The aether and the imponderable fluids were now widely understood to provide a new theoretical framework for natural philosophy. Pneumatic chemistry, a field that owed much to Cavendish's work, was just then acquiring a caloric theory of gases, according to which the particles of gases are surrounded by a repellent fiery matter, an idea foreign to Cavendish's way of thinking. By explaining the phenomena of heat by forces of attraction and repulsion between the particles of ordinary matter, Cavendish showed that Newton's main direction in natural philosophy could accommodate recent experimental work. "Heat" addressed more than a theory of heat. Doubtless for Cavendish an orientation of thinking in natural philosophy was also at stake.

Publication

Cavendish, we know, wrote "Heat" with the intention of publishing it. He referred to the first draft as the "foul copy," appending to it several pages of additions and alterations. In the second draft, he referred to the "text," for which he provided an apparatus of footnotes, and he planned yet another writing in which certain paragraphs would be changed. He called the whole a "paper." In scope and ambition, it compared with his published paper on electrical theory.

For the sake of argument, let us suppose that Cavendish had submitted his paper to the Royal Society. A slightly abbreviated version would have been read at a meeting, and the entire paper would have been considered by the papers committee for publication in the *Philosophical Transactions*. In a general way, the members would have found in Cavendish's paper much to admire. They would certainly have commended it for adhering to widely-held objectives of natural philosophy. First, as we have noted, it addressed a persisting problem in the foundations of natural philosophy, the cause of heat. Second, because heat was the common currency of natural philosophy, its theory necessarily connected diverse physical subjects, contributing to the overall coherence of the field. Third, it laid the groundwork for the next stage of research on heat, the discovery of new facts. Fourth, the theory was well constructed, which meant that it was comprehensive, embracing most of the established regularities of heat; that it was exact, precise, quantitative, and compatible with the instruments of experimental physics; that it was mathematical; that it reflected the simplicity of nature in the economy of its premises; that it proceeded from the laws of nature; that its reasoning was unobjectionable; that it was mechanical; and that in some acceptable meaning of the term, it was Newtonian. Finally, should its hypothesis be confirmed, as a mathematical branch of natural philosophy, the theory of heat would join a select company of theories: "Celestial Physics," the "first in dignity of all inquiries into Nature whatever,"[35] and, close behind, the mathematical theories of mechanics and optics.

With that going for it, why did Cavendish not follow through with his intention? If not exactly a reason for withholding publication, there is a consideration: given the history of vis viva in Netownian polemics, to advance a theory called Newton's theory, and to base it on Leibniz' vis viva was somewhat provocative. Cavendish was confident, but how bold he was is harder to say.

Maybe Cavendish was disappointed with the finished product. In a discussion of the heat of electrical discharge, he noted, "This must be examined,"[36] and he conceded that he could not explain why bodies always expand with heat rather than sometimes contract,[37] but he found neither these nor any other facts inconsistent with the theory. In its mathematical development, the theory of heat fell short of his electrical theory, and the theory did not give clear experimental directions. Just how much weight he gave to these limitations we have no way of knowing for certain, but probably not much, since he would have been aware of them at the outset. Perhaps he simply dropped the work to take up another of his many interests; there were plenty of other researches he did not bring to publication. An argument against this is the his obvious intention from the beginning to publish the theory. Perhaps he lost confidence in the mechanical theory of heat in general. At about the same time as "Heat," he abandoned another theory, the phlogiston theory of chemistry; he was fully capable of doing it again, only there is no evidence that he did. In the case of heat there was rumor he did so, but there is no evidence for it. Lavoisier's chemistry listed caloric, the matter of fire, among the elements, and although we have no reason to think that Cavendish accepted caloric along with oxygen, it is not inconceivable. We have no statements by him on the nature of heat later than what appears in "Heat."

Or maybe he decided it was senseless to publish on a theory that looked to be on its way out. Supporters of the material theory were unlikely to find in his paper compelling reasons to change their mind. They were given no new experiments, and as for hoping to persuade them on theoretical grounds, Cavendish knew his audience. A few years after Cavendish's paper, a student of Joseph Black's, John Leslie, said that the interpretation of heat as the vibrations of particles of bodies still had some adherents, but they were sadly misguided; in addition to the "insurmountable objections" to this interpretation, it was "vague and undefined," a "shapeless hypothesis," "merely nugatory," and "explains nothing."[38] An advocate of the fluid theory of heat near the close of the century called the mechanical theory of heat one of the "inconceivable and incredible mysteries that philosophy propounds."[39] Apart from the notorious difficulty of shaking firmly held convictions, Cavendish's paper faced the even greater difficulty of finding readers at all. Although the mathematics he used was not unusual, his paper could have been read with full appreciation only by natural philosophers who were at home with mathematical arguments, who were not many. His electrical theory had found precious few readers for similar reasons. The mechanical theory of heat had enormous potential, but Cavendish's contemporaries could not have known it. This being said, I doubt that Cavendish was deterred by obstacles of this sort. He had an audience, small to be sure, but choice. His like-minded friend Thomas Young observed that the "most sober reasoners of the present" held the vibration theory of heat.[40]

Finally, there is the question of priority. When a few years before "Heat" Michell asked Cavendish not to reveal a scientific paper he had sent him until it appeared in print, Cavendish protested that "the surest way of securing merit to the author is to let it be known as soon as possible & those who act otherwise commonly find themselves forestalled by others." Instead of wishing him to keep it secret, Michell should have him "shew the paper to as many of your friends as are desirous of reading it"; Michell agreed with Cavendish.[41] In light of this sound advice, it is possible that Cavendish did not publish "Heat" because he had delayed too long. That possibility is unlikely, however, because in Cavendish's lifetime no work like his appeared in print. After Bernoulli's brief discussion, the first publication on the kinetic theory of gases came out only after Cavendish had been dead for six years, and it did not identify heat with vis viva, but with momentum. Experiments on the mechanical equivalent of heat began to appear only in the 1840s, and publications on the mechanical theory of heat only in the 1850s.

It would seem that Cavendish took to the grave his reasons for not publishing his theory of heat. We can assume that they were related to the reasons why he wrote it in the first place, concerning which we can only make educated guesses. In the final analysis, we suppose that his argument for the Newtonian view of heat somehow fell short in his estimation.

Newton's Theory

We may refer to the author of "Heat" as "Newtonian." Historians of science warn of the vagueness of the label, as they do against inferring an exaggerated influence from the ubiquity of contemporary references to Newton. The point is worth making, but it should also be noted that Newton was no more removed in time from Cavendish than Einstein is from us, and as Einstein is ever-present in certain lines of work in physics today, so was Newton then. Cavendish pointedly called his theory of heat "Newton's theory."

"When a theory has been proposed by Sr I[saac] N[ewton]," Cavendish wrote in "Heat," and it agrees with experience, it is to be accepted.[42] To no other authority did he give an endorsement like this. If there is a doctrinaire element in Cavendish's thinking, it would be vindicated in the next century with the success of the theory of heat as motion.

Borrowed authority was unlikely to have been a motive in Cavendish's case. If anything, the converse was more likely. Cavendish's authority would have validated the Newtonian philosophy, which by then could use support. He would have been familiar with the opinion that Newton had discovered the laws of ordinary bodies but not those of elementary fire, the cause of heat.[43] "Heat" was Cavendish's rejoinder. With it, he could argue that Newton had not overlooked a cause but rather had advocated the true cause of heat, only waiting to be developed.

Wanting concepts of specific and latent heats, Newton could not have made a satisfactory mathematical theory of heat. What he could do, he did: He made observations on heat, noted its intimate association with diverse phenomena throughout natural philosophy, and left behind a wealth of suggestions for further inquiry. In speaking of Newton's theory in connection with his own, Cavendish acknowledged a general indebtedness to Newton.

Cavendish's theory was Newtonian in another way. The analogy of nature had led Newton to expect that the minutest bodies of the world, particles, would be understood by the same principles he had used to understand the greatest bodies, the sun and planets. Early British Newtonians were believers, persuaded that mathematical laws of short-range forces between particles, or corpuscles, were capable of explaining physical and chemical processes. The expectation died hard. Looking back, a late-eighteenth-century observer wondered why so little had been accomplished, why there had been so little mathematical work on corpuscular phenomena in Newton's time. Newton, after all, had provided the necessary mathematics together with "a new system of mechanics, particularly fitted for this research, and had demonstrated its competency by the most successful examination of the great movements of the universe," and in the queries of his *Opticks* he had given "very encouraging analogies, which seemed to admit the same manner of treatment in the study of the corpuscular phenomena." "Nothing however was done that was of any service," nor had much been done in the intervening hundred years. The reason seemed to have been the "immense difficulty of the task," which had deterred even Newton.[44] Cavendish took up the task in "Heat," with a theory based on a known law of force, the "force of moving bodies." His use of vis viva as the measure of the attractive and repulsive forces between particles freed him from the need to know the mathematical laws of the short-range forces acting in the interior of matter, ignorance of which had obstructed mathematical progress in the corpuscular philosophy in the past. With his mathematical theory of the motions constituting heat, Cavendish fulfilled an old expectation, and at the same time he looked to remove the loose reasoning that passed for Newtonian natural philosophy in the centenary year of the *Principia*.

To make a workable Newtonian theory of heat, Cavendish needed more than Newton gave him. Important as Leibnizian vis viva was for his purposes, that did not give him his theory either. He needed the full resources of natural philosophy to write "Heat."

Natural Philosopher

Part 1 of this book discussed natural philosophers in general. Let us close by returning to the subject, now with our specific example, Cavendish. James Hutton distinguished between "science" and "philosophy," between possessing facts and understanding their place in natural knowledge. Conceding that science cannot advance one step without experiment, Hutton said that unless experiment is guided by philosophy, science can produce only endless collections of facts. "The disposing of one fact, that is, the putting it into its proper place in science, or the general order of our knowledge, is doing more for natural philosophy, than a thousand experiments made without that order of connection or relation which is to inform the understanding."[45] If we accept Hutton's characterization of the work of a natural philosopher, then we would exclude Cavendish, as he is portrayed in George Wilson's *The Life of the Honourable Henry Cavendish*, from the ranks of natural philosophers. Cavendish's universe, Wilson declared, consisted "*solely* of a multitude of objects which could be weighed, numbered, and measured; his brain was but a "calculating engine."[46] Wilson's judgment has been uncritically repeated ever since, but he was fundamentally in error about his subject. Cavendish's "universe" contained everything that natural philosophy addressed, and his "brain" was, among other things, a marvel of theoretical perspicuity.

The "Business of natural Philosophy," Hugh Hamilton wrote, is "to reduce as many Phaenomena as may be to some general well-known Cause."[47] That is a good description of what Cavendish did. In all three of his major original lines of research, chemistry, electricity, and heat, he held theories of the cause of the phenomena. His chemical researches were guided by the phlogiston theory, discussed in his earliest chemical writings. The starting point of his electrical researches was his theory of an electric fluid. As we know, he held the Newtonian theory of heat. Throughout his career, he worked from these theoretical ideas—he gave up phlogiston only after his last publication in chemistry—modifying them as needed, and studying the phenomena in question experimentally. Cavendish's goal was not, as has been suggested, to measure and calculate for their own sake; that much is evident from "Heat." His paper contains no measurements of heat, and for the most part the subject of heat did not yet lend itself to extended calculations. Rather "Heat" brought Hutton's "order of connection" to the facts of heat, "to inform the understanding," the point of natural philosophy.

From the perspective of the history of science, Cavendish's theory of heat seems decidedly ahead of schedule, and as such it is yet another contribution to his peculiar reputation for having anticipated work by later physicists. Historical studies of theoretical physics from the middle of the nineteenth century emphasize a number of developments.[48] One of these is the emerging concept of energy together with its law of conservation, which William Thomson called the greatest advance in physical science "since the days of Newton," leading to a conceptual unification of physics.[49] Cavendish gave a precise version of that law in "Heat." Other developments were the rejection of the physics of imponderable fluids and its replacement by the mechanics of molecules, the use of mechanical analogies, the use of hypotheses, the theoretical guidance of experiment, the mathematical formulation of theory, and a concern with qualities that make a good physical theory. These developments all make their appearance in "Heat." At the very least, upon reading Cavendish's theory of heat, we can say that the strong direction of theoretical physics from the middle of the nineteenth century had substantial antecedents in the natural philosophy of the late eighteenth century.

Henry Cavendish's Manuscript on the Mechanical Theory of Heat

EDITORIAL NOTE

After our excursion through the history of science, we are ready to enter a theoretical work in the age of natural philosophy. Proceeding from the most secure theory of physics, mechanics, thoroughly grounded in the new quantitative laws of experimental physics, fomulated mathematically, and explicit on the philosophical issues of physical theory, Henry Cavendish's theory of heat offers a perspective on the world of eighteenth-century natural philosophy.

"Heat" carries no date.[1] It was certainly written after 1783,[2] for as we have seen, in that year Cavendish still rejected Joseph Black's term "latent heat," which he used without qualification in "Heat." Cavendish cited Joseph Priestley's book on the history of optics, but that appeared early, in 1772. He cited the names, but not the publications, of C. W. Scheele and H. B. de Saussure for their researches on radiant heat. The reference in "Heat" shows his familiarity with Scheele's only book, which appeared in English translation in 1780.[3] Cavendish's reference to Saussure was no doubt to the second volume of his travels in the Alps, which came out in 1786.[4] The absence of citations of work done in the 1790s may be taken as indirect evidence that the manuscript was written before then.[5] After Abraham Bennet's repetition and criticism of John Michell's experiment on the momentum of light in 1792, it is unlikely that Cavendish would have given this experiment the prominence he did in "Heat." A subscriber to Bennet's book on electricity, Cavendish evidently respected his work.[6]

Cavendish's paper on the mechanical theory of heat is written in ink on quarto sheets and wrapped in a folder labeled, in his hand, "Heat." The manuscript exists in three parts, which correspond to three stages of the work. The first, a two-page sketch, is a partial list of topics taken up in the paper, noting difficult points of the theory. We think that it is a preliminary list belonging to the planning stage, but it could have been written later, during or after the writing of the paper. The foul copy of the complete paper, which lays out the argument of the theory, consists of forty pages of text and footnotes with four pages of additions and alterations. The revised copy tightens the argument, sets apart definitions, numbers discussions, supplies punctuation and reorganizes materials. It consists of forty-three pages of text and footnotes, one page of diagrams with an accompanying page of explanations and one page of additions and alterations. The sketch and both versions of the paper are included in this edition. There are three reasons for including everything: First, the manuscript is not long. Second, a complete manuscript is better than a partial manuscript, since readers come to it with different questions. Third, it brings us as close as we are likely to get to the thinking processes of this natural philosopher. Most readers consulting this edition will go directly to the fair copy beginning on p. 176.

What Newton and other writers called "scholia," discussions that were inessential to the argument, Cavendish relegated to footnotes in the paper. Following his directions, in this edition, his footnotes are designated by asterisks and other signs.

Rearrangements and insertions of paragraphs, and minor rewordings throughout, are made without comment, as directed by the author. This gives the foul copy a far more finished appearance than the first draft conveys, but without it the paper is difficult to read. Possibly significant deletions are enclosed within angled brackets (< >) and run into the text or given in editorial footnotes. All editorial additions and comments are enclosed in square brackets ([ ]). The number of dashes within a pair of brackets indicates the number of illegible characters, five dashes standing for the maximum number. Unusual spellings or misspellings are allowed to stand, and missing punctuation is indicated by an extra space between words. Cavendish's drawings have been redrawn for this book. For convenience of reference, the drawings are located in the text where they are mentioned and are numbered accordingly.

[PRELIMINARY SKETCH OF THE MANUSCRIPT]

[Heat]

Heat from action of ⊙^{s} light   no heat when light is reflected or refracted but only when the rays are stopt or [-----]   heat which glass & polishd metals receive when exposed to fire   calcuation of vis viva of ⊙^{s} rays & D^o required to commun. given quant. heat

Heat by emission of light   the light commonly impelled by repulsion of large portions of matter <Whether in flame the rays emitted can be impelled by repulsion of large portions> but quere whether this can be the case in flame

Heat by electricity

Heat by friction & hammering   Whether they can give suffic. vis viva   Perhaps may where much force is concentrated in small space as in boring holes &^{ct} but as friction is not produced without tearing the greatest part of heat produced thereby is likely to be owing to other cause   Whether all kinds of force applied should give any vis viva to a body or only suffic. quick motions

What is the cause of friction & want of elasticity   whether it is not always owing to tearing off of particles or altering their arrangement   It should nat. seem that if there are no solid particles or if they do not touch each other that though it might require some force to first put the bodies touching each other in motion yet it should require none to keep them in mot.   exper. whether friction is as much diminished by oil & grease as the heat is

<Do not know> Cannot explain why the motion of the particles should cause a body to expand but as the action of the particles on one another can hardly be the same as if they were at rest it is reasonable to expect that their arrangement & distance should be alterd thereby & consequently it is not extraordinary that the body should be made to expand thereby

[FOUL COPY OF THE MANUSCRIPT]

[Heat]

The effect which a body in motion can produce in mechanical purposes is as the weight of the body multiplied by the square of its velocity or as its weight multiplied by the height from which it must fall by its weight to acquire that velocity   this force I shall call its vis viva & for shortness if the weight of the body is g gra.   & its velocity is such as it would acquire by falling f feet I shall call its vis viva g gra × f feet & so on

In computing the vis viva of a system of bodies   The vis viva of each body must be taken separately by multiplying its weight by the square of its whole velocity without regard to the direction of that velocity & taking the sum

Pr. 1)   Let there be a system of bodies which are moving in such a manner that their center of gravity is at rest & let the vis viva of that be called S   Let now each body receive a new velocity v in a given direction so that the bodies will move in respect of each other in exactly the same manner as before but their center of gravity will move with the velocity v instead of being at rest as before & let the weight of the whole system multiplied by the square of v be called T   then will the whole vis viva of the system be S + T

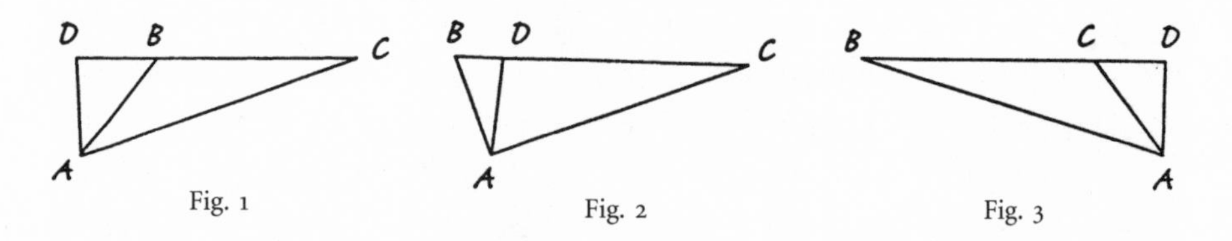

Fig. 1   Fig. 2   Fig. 3

For let B be any body of this system   suppose that before it received its new velocity it was moving with the velocity AB in the direction AB that is from A to B & let the new velocity given to it be BC & in the direction BC   therefore after that it will move with the velocity AC in the direction AC   therefore if the weight of the body is called w its vis viva will be w × AC^2 = w × AB^2 +

w × BC^2 + w × 2BD × BC observing that if D is on the same side of B as C is as in fig. 2 & 3 then BD must be considerd as negative But the sum of all the quantities w × AB^2 = S & the sum of all the quantities w × BC^2 = T & the sum of all the quantities w × 2BD × BC is equal to the weight of the whole system multiplied by 2^{ce} the product of the velocity which the center of gravity had in the direction CB before the application of the new velocity & of the velocity v which by the supposition is nothing & consequently the whole vis viva of the system after the application of the new velocity will be S + T

Coroll.) Let there be a body whose particles are in motion but in such manner that the body may be divided into a vast number of very small portions whose centers of gravity are at rest & let the vis viva of this body be call S as before This may be called the invisible vis viva of the body as it consists only in the internal motion of its invisible particles Let now a motion be given to this body either simply progressive or rotatory or both but in such manner as not to alter the motion of the particles in respect of each other Let the vis viva which the body would acquire in consequence of this motion supposing the internal motion of the particles to be stopt be called T This may be called the visible vis viva of the body Then the whole vis viva of the body or the sum of the vires vivae of each particle does not differ sensibly from S + T or is equal to the sum of the visible & invisible vires vivae

For as the abovementiond portions are small their rotatory velocity must be small in respect of the velocity of their center of gravity & consequently the vis viva of each small portion is equal to the vis viva of its center of gravity & it was before shewn that the whole vis viva of a

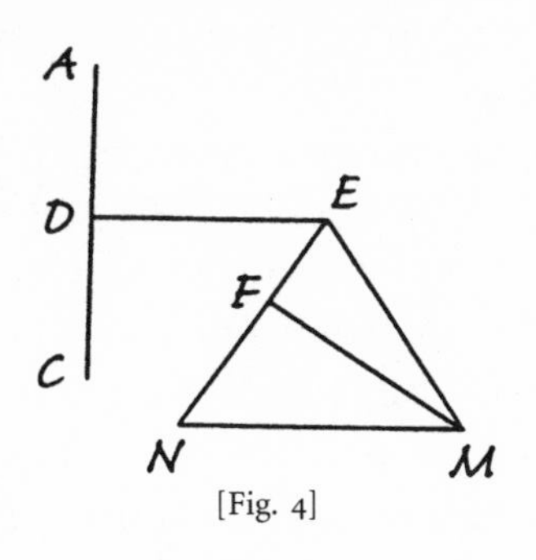

[Fig. 4]

Defin

Let there be any body whose particles are in motion Let C be its center of gravity & CA any line drawn through it which we will call its axis Let E be any particle & ED a perpendicular from it to the axis & let EF be the rotatory velocity of that point that is let it be equal to that part of its velocity which is performed in the line EF perpend. to DE & in a plane perpend. to the axis & let the weight of E be multiplied by the product of DE & EF & let the same thing be done by every other particle observing that if the rotatory velocity is in the contrary direction to that of E id est such as to make it revolve the contrary way its product must be considerd as negative then if the sum of all these products is nothing the body is said to have no rotatory vis viva on the axis AC & if it has no rotatory vis viva on any other axis it is said to have no rotatory vis viva

Prop. 2 Let the particles of the body be in motion but in such manner as to have no rotatory vis viva & [such] that its center of gravity is at rest & let its vis viva be S Let now a rotatory velocity be given to each particle proportional to its distance from the axis so that the body will acquire a rotatory vis viva with[out] altering the relative motions of the particles in respect of each other & let T be equal to what would be the vis viva of the body if its particles had only this rotatory motion Then will the whole vis viva of the body be S + T

For take any particle E let its weight = w suppose that before the new motion given to it it was moving with the velocity ME in the direction ME consequently if EN is drawn in the direction of its rotatory velocity & MF is drawn perpend. to EN EF was its rotary velocity Let EN be the new rotatory velocity given to it which by the supposition is proportional to ED then its vis viva before the application of this new motion was $w \times ME^2$ & after it is $w \times MN^2$ which equals $w \times ME^2 + w \times EN^2 - 2w \times EN \times EF$ But the sum of all the quantities $w \times ME^2 = S$ & the sum of all the quantities $w \times EN^2 = T$ & the sum of the quantities $2w \times EN \times EF$ is by the supposition equal to nothing & therefore the whole vis viva of the body is S + T

Coroll) The quantity S may properly be called the invisible vis viva of the body as it arises only from the motion of the invisible particles of the body among each other & in like manner the quantity T may be called the visible vis viva Now it appears from the 2 foregoing propositions that whether the body has a progressive or rotatory vis viva or both together its total vis viva is equal to the sum of its visible & invisible vis viva & if the visible vis viva is diminished without altering the total vis viva the invisible vis viva must be as much increased

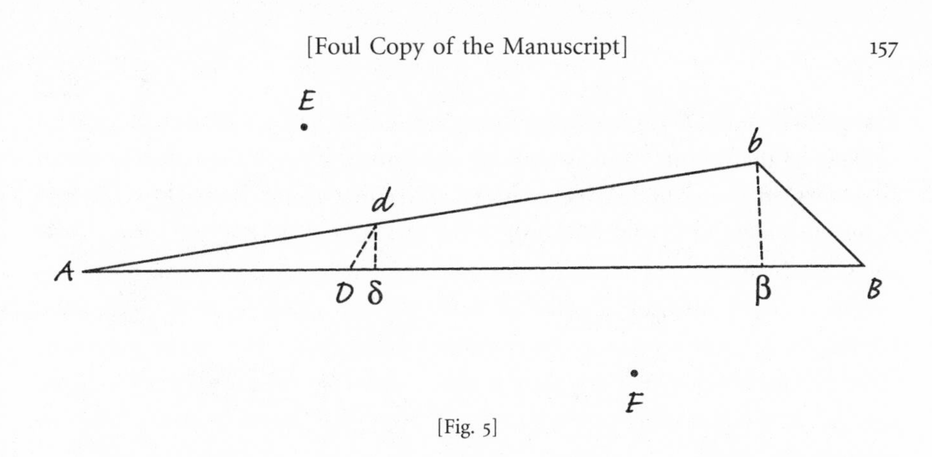

[Fig. 5]

Let[2] the particles B D E & F attract or repel each other in such manner that the attraction or repulsion of any 2 particles as B & D on each other shall be every where the same at the same distance from each other however different at different distances Let B D E & F be the present position of these particles & suppose that at the end of a given short time B & D are removed to b & d Let BD & bd continued meet in A & let this time be so small that the angle bAB shall be infin. small & with the center [A] draw the arches dδ and bβ. Bβ & Dδ is the space by which B & D approach nearer together observing that if β is placed further from D than B is so that the motion of B is such that B recedes from D Bβ must be considerd as neg. & the like must be observed with regard to Dδ Let now $\dot{B}$ be the increase of vis viva which B receives in that time by the action of the other particles observing that if B receives a diminution of its vis viva $\dot{B}$ must be considered as neg. & $\dot{\overline{bd}}$ that which it would receive in the [same] time by the action of D alone supposing it to be confined to move in the direction Bb but without being acted on by E or F Let $\dot{\overline{\beta d}}$ be the vis viva which B would receive in falling through the space Bβ by the attraction of D & let $\dot{\overline{BD}}$ be the increase of vis viva which either B or D would acquire by falling through Bβ + Dδ or BD − bd by their mutual attraction* observing that it must be considerd as before that $\dot{\overline{BD}}$ must be pos. or neg. according as Bβ + Dδ & the action of B & D are [aff]ected both by the same sign or by contrary signs & let increments of vis viva of the other particles D E & F be

*The vis viva which D will acquire in falling through any space by the attraction of itself & [B] to each other is the same that B will acquire by the same cause in moving over the same space & the increase of vis viva which a body already in motion acquires in falling through a given space by a given attraction is equal to the whole vis viva which it would acquire in falling from rest through the same space

expressed in the same manner by their respective letters Then the increase of vis viva which B receives in the above mentiond short time in consequence of the attraction of D or $\overline{bd}^{\cdot}$ is equal to the vis viva which it would receive in falling through Bβ by the attraction of D or to $\overline{\beta d}^{\cdot}$ that [is] $\overline{bd}^{\cdot} - \overline{\beta d}^{\cdot} = 0$ & therefore

$$\begin{aligned}
&\overline{bd}^{\cdot} + \overline{be}^{\cdot} + \overline{bf}^{\cdot} - \overline{\beta d}^{\cdot} - \overline{\beta e}^{\cdot} - \overline{\beta f}^{\cdot} = 0 \\
&\overline{db}^{\cdot} + \overline{de}^{\cdot} + \overline{df}^{\cdot} - \overline{\delta b}^{\cdot} - \overline{\delta e}^{\cdot} - \overline{\delta f}^{\cdot} = 0 \\
&\overline{eb}^{\cdot} + \overline{ed}^{\cdot} + \overline{ef}^{\cdot} - \overline{\varepsilon b}^{\cdot} - \overline{\varepsilon d}^{\cdot} - \overline{\varepsilon f}^{\cdot} = 0 \\
&\overline{fb}^{\cdot} + \overline{fd}^{\cdot} + \overline{fe}^{\cdot} - \overline{\phi b}^{\cdot} - \overline{\phi d}^{\cdot} - \overline{\phi e}^{\cdot} = 0
\end{aligned}$$

But $\overline{bd}^{\cdot} + \overline{be}^{\cdot} + \overline{bf}^{\cdot} = \dot{B}$ & $\overline{\beta d}^{\cdot} + \overline{\delta b}^{\cdot} = \overline{BD}^{\cdot}$ & consequently $\dot{B} + \dot{D} + \dot{E} + \dot{F} - \overline{\beta d}^{\cdot} - \overline{\delta b}^{\cdot} - \overline{\beta e}^{\cdot} - \overline{\varepsilon b}^{\cdot} - \overline{\beta f}^{\cdot} - \overline{\phi b}^{\cdot} - \overline{\delta e}^{\cdot} - \overline{\varepsilon d}^{\cdot} - \overline{\delta f}^{\cdot} - \overline{\phi d}^{\cdot} - \overline{\varepsilon \phi}^{\cdot} - \overline{\phi e}^{\cdot} = \dot{B} + \dot{D} + \dot{E} + \dot{F} - \overline{BD}^{\cdot} - \overline{BE}^{\cdot} - \overline{BF}^{\cdot} - \overline{DE}^{\cdot} - \overline{DF}^{\cdot} - \overline{EF}^{\cdot} = 0$ that is the sum of the vires vivae which each particle receives by their mutual attraction [minus] the sum of the vires vivae which each particle would receive by the attraction of each other particle in moving through ½ the distance by which they approach each other remains unalterd

Coroll.) Let there be a body consisting of any number of particles attracting or repelling each other as above Let the vis viva which each particle would acquire in falling by the attraction or repulsion of each particle from an infinite distance to its actual distance from that particle be computed separately observing that if the 2 particles repel each other this vis viva must be considerd as negative & let ½ the sum of all these vires vivae be called S Let these particles be supposed in motion & let the sum of their actual vires vivae be called s then s − S remains constant & is not alterd by their motions among each other

Heat I suppose consists in the internal motion of the particles of which bodies consist but though these particles are in constant motion yet they are so far retaind in their place by their mutual attractions & repulsions that while the nature of the body remains unchanged the greatest distance to which they can be removed by these vibrations from their original situation must be excessively small It follows therefore that though the values of S & s will be perpetually altering & will never remain exactly the same even for an instant yet unless they are disturbed by some external cause they can never vary sensibly from their present value or in other words its heat will remain constantly the same But if by any means the size of the body or the arrangement of its particles is alterd then as the value of S can hardly escape being alterd thereby the value of s must be alterd as much or in other words the heat of the body will be alterd

On the communication of heat from one body to another

By the sensible heat of a body I mean its heat as shewn by a thermometer

By the quantity of latent heat in a body I mean the value of −S   By the quantity of active heat I mean the value of s & consequently s − S is the quantity of total heat   By the capacity of a body for heat I mean the quantity of total heat which must be commun. to a given weight of that body in order to produce in it a given increase of sensible heat   It is evident that we have no means of judging of the quantity of either the latent or active heat in a body or the alterations which they undergo but the alterations of the total heat in a body may be measured

Let 2 bodies A & B precisely of the same nature be put in contact & let the value of s in A be greater in proportion to its size in A than in B   then it seems clear that the particles of A will communicate part of their motion to those of B till the particles of both bodies come to vibrate with the same velocity that is till the value of s is the same in proportion to their size in both provided neither are in contact with any 3rd body to which they can communicate or from which they can receive motion

But if the bodies A & B are of different natures it seems likely that the value of s should never become the same in proportion to its bulk in both though they are kept ever so long in contact   For example if the vibrating particles of B are greater than those of A it is likely that the value of s will always be less in proportion to its bulk in B than in A*

But though the value of s is not the same in proportion to the bulk in A as in B yet if a 3rd body C is put in contact with A & a 4th body D precisely of the same nature as C is put in contact with B it seems reasonable to suppose that the 4 bodies would communicate motion from one to the other till the value of s became the same in proportion to the bulk in C as in D or till D had the same quantity of vis viva in proportion to its size as C so that A & B would still shew the same degree of heat by the thermometer   Thus if we suppose that when 2 equal bodies remain long enough in contact their vires vivae will be to each other inversely as the size of their particles & that the size of the particles in A B & D are as α β & δ then as C is supposed to be of the same nature as D its vis viva will be to that of A as α:δ & the vis viva of B will

*Though I can not give an absolute demonstration of this yet it seems likely from this circumstance that if 2 bodies of different sizes are at rest & recede from each other by their mutual repulsion the larger body will acquire thereby a less vis viva than the smaller

be to that of A as $\alpha:\beta$ & the vis viva of D will be to that of B as $\beta:\delta$ & conseq. its vis viva will be to that of A as α to δ and conseq. its vis viva will be equal to that of C

From what has been said it appears that when 2 bodies unequally heated however different their natures may be are brought in contact the hotter body must commun. heat to the other till they both acquire the same sensible heat provided neither of them are in contact with other bodies which can carry off from or communicate heat to it & 2ndly the quantity of total heat which one body parts with must be equal to that which the other acquires 3rdly some bodies may require a greater addition of total heat to produce in them a given increase of sensible heat than others & this from two causes first that some bodies may require a greater addition of active heat than others in order to produce a given increase of sensible heat as was before said in art & 2ndly because in all bodies the alteration of sensible heat can hardly help being attended with an alteration of the quantity of latent heat for as the bulk of all or at least almost all bodies is increased by heat the distance of their particles must be alterd which can hardly fail of being attended by an alteration of the value of S that is of their latent heat & that alteration can hardly fail of being greater in some bodies than in others

On the heat <& cold> produced by chemical mixtures

It seems a natural conseq[uence] of this theory, that the mixture of 2 substances which have a chymical affinity should commonly be attended by an alteration of sensible heat for as the arrangement of the particles must be alterd thereby the quantity of latent heat can hardly fail of being alterd & moreover it is very possible that the quantity of active heat necessary to produce a given sensible heat may also be alterd from both which causes the quantity of total heat necessary to produce a given sensible heat will be alterd & consequently as the quantity of total heat cannot be alterd by the mixture the sensible heat must be alterd

What is here said is applicable to the case of the cold produced by the changing of bodies from a solid to a fluid form &ct for as bodies in a fluid form must in all probability <require> a different quantity of heat from what they do in a solid form it will require a great addition or substraction of total heat to reduce them from one form to another though their sensible heat suffers no sensible alteration It must be observed however that I do not by this mean to account for the circumstance that cold is always produced by the changing from a solid to a fluid form & vice versa but only that it should require a considerable alteration of total heat to produce this change

On heat from the impulse of light

If the body A is at rest & another body B moves towards it & when it comes near it is repelled by it & turned out of its course or reflected then if A is not much bigger than B B will lose thereby a considerable part of its vis viva & A will acquire as much but if A is excessively bigger than B B will lose but an excessively small part of its vis viva & will communicate excessively little to A but in all cases A will acquire as much vis viva as B loses

There can be no doubt that light consists of excessively small particles emitted with excessive velocity from the luminous body & it has been sufficiently proved that when light is reflected from any body the particles of light are not reflected by impinging against any solid particles or even by the repulsion of a few small particles only but by the joint repulsion of a quantity of matter infinitely greater in quantity than the particles themselves <It follows therefore that a body can receive no heat from light reflected from it or refracted through it but only from that part of the light which is stopt in it & that the total heat produced thereby is equal to the vis viva of the particles stopt in it> so that they can lose no sensible part of their vis viva thereby & can communicate no sensible quantity to the body & the same thing takes place when light passes through a transparent body & is refracted but when light enters into a body & is stopped there & does not escape then the particles will be continually reflected backwards & forwards till they at last come to vibrate with no greater velocity than the particles of the body it self so that their vis viva will be equally distributed between them & the body

In order to find the vis viva of all the $\odot^{s}$ light which falls in 1" on a surface of 1½ sq. feet let the weight of the light which falls on that surface in that time be called w & let the velocity of light be v inches per " a body will by falling by its gravity through 16 feet acquire a velocity of 32 × 12 inches per " & therefore the vis viva of the light which falls in 1" on 1 ½ sq. feet = w drawn into $16 \times \left(\dfrac{v^2}{32 \times 12}\right)^2$ feet Let now this light be received on a plate exposed perpendicularly to it whose weight = p then if this plate reflects all the light it will by its impulse receive in 1" a velocity of $2v \times \dfrac{w}{p}$ inches per " but if it absorbs all the light it will receive only ½ that velocity Now according to an experiment of Mr Mitchells[3] related by Dr Priestley it seemed that the impulse[4] of the light falling from the ⊙ on 1½ sq. feet of surface was sufficient to give in 1" of time to a plate weighing 10 gra a velocity of 1 inch per " & therefore if we suppose that the plate reflected all the light 2wv/10 gra must equal 1 or wv must = 5 gra & consequently the vis viva of the ⊙s light which falls in 1" on a surface of 1½ feet must equal 5 gra $\times \dfrac{v}{64 \times 12 \times 12}$ feet which as[5] v = 12000.000.000 equals 6500.000 grains × 1 foot but if we suppose that plate absorbed all the light then the vis viva of light is 2ce as great If a horse working in a mill raises 100 £ at the rate of 3 miles an hour or 4½ feet a " the labour of a horse in 1" is sufficient to produce a vis viva of 700.000 gra × 4½ feet or 3150.000 gra. × 1 foot so that the vis viva of the ⊙s light falling on 1½ sq feet is equal to the labour of more than 2 horses*

Exper to determine the quantity of vis viva necessary to give a given increase [of] sensible [heat] to a given body by alternately exposing a body to the ⊙ & shading it & thereby to give a guess at the velocity with which the particles of a body vibrate supposing that the total heat of a body heated to 1000° is double its heat at 0°

It was said that bodies are not [heated] by rays refracted & transmitted through them but only by rays absorbed by the body but yet it has been found that a plate of glass is much more heated by the fire (& I believe by the ⊙) than a plate of polished metal though the metal plate absorbs more light than the glass one which at first sight seems contradictory but the reason of this is easily explaind by the observation of Scheele[6] as he has satisfactorily shewn that hot bodies emit not only rays of light but also emit other particles not capable of exciting the sensation of light in our eyes but which are yet capable of exciting heat & which may therefore be called rays of heat & that these rays of heat are reflected by polished metals but are not reflected nor transmitted by glass it therefore is not extraordinary that the glass should be heated more than the metal as it is heated by the rays of heat The rays of the ⊙ seem to contain a less proportion of rays of heat than those of a fire & Mr Saussure[7] has found that bodies emit rays of heat though [not] near hot enough to emit rays of

*If it was possible to make a wheel with float boards like a water wheel so as to move with the velocity of light without suffering any resistance from friction & the resistance of the air & as much of the ⊙s light as falls on 1½ sq. feet was thrown on one side of this wheel it would actually do as much work for any mechanical purpose as 1 horse if the float boards absorbed all the light which fell on them or as 2 horses if they reflected it all

light & even though not hotter than boiling water so that it should seem that the more intensely a body is heated the less proportion of rays of heat it emitts

It is uncertain with what velocity the rays of heat are emitted if they are emitted with less velocity than those of light it make[s] the vis viva of the ⊙s rays less than according to the above mentiond computation but the error arising from thence cannot be very great as in the ⊙s rays the heat produced by the rays of light seem to be not less than that by the rays of heat since a burning lens seems to burn as well as a metallic mirror of the same size

From what has been said it follows that all bodies exposed to the ⊙s rays ought to receive an equal addition of total heat by it provided they are equally dark colourd except so far as depends on some of them absorbing more of the rays of heat than others but the sensible heat which they will acquire depends also on their capacities for heat & the ease with which they transmit heat to the air & other bodies in contact with them that is to the swiftness with which they lose the heat communicated to them (Exper to try this) Expose thermometers whose bulbs are coated with various dark & equally dark colourd substances alternately to the ⊙ & shade & see whether they receive the same increase of heat in the same time observing that the coatings must be so thin that the total heat required to heat them shall be small in respect of that required to heat the bulb If they do not receive nearly the same increase of heat expose them again to the ⊙s light transmitted through a flat glass so that most of the rays of heat shall be absorbed before they fall on the thermom. If they now approach much nearer towards acquiring the same heat it will be a proof that if it was not for the unequal absorbtion of those rays of heat they would have received the same quantity of total heat

N.B As some bodies are much affected by light independent of the heat they receive by it they are evidently improper substances to coat the thermometers with as their latent heat may be much affected by the light

If it should prove that different bodies do not receive the same total heat from the ⊙s light it would be difficult to reconcile with this hypothesis But then it seems as difficult to reconcile it with the supposition of heat being a material substance except that as those hypotheses are less capable of being brought to the test of strict reasoning it is easier for those gentlemen to find loop holes to escape by

Whether heat is produced by the emission of light

If the light receivs its velocity from the repulsion of a quantity of matter vastly superior in weight to that of the particles emitted it is plain that no sensible heat can be produced by the emission of light If on the contrary the weight of this quantity of matter is not vastly superior to that of the particles of light themselves some heat will be generated It is much more likely that the former supposition is the truest & consequently that no sensible heat is generated by the emission of light nor does it seem possible to ascertain by experiment whether heat <is generated by the emission of light>

On the heat produced by friction & hammering

Whenever one body rubs against another or whenever 2 unelastic bodies strike against each other it is natural to suppose that the force lost thereby will be spent in communicating vis viva to the particles of the bodies though on the other hand if the velocity with which the bodies rub or impinge against each other is less than that with which their particles vibrate one does not readily see why that rubbing or impinging should increase the vibrating velocity of the particles but which ever of these is the case it seems likely from what was said of the quantity of vis viva requisite to produce a given alteration of sensible heat in bodies that no perceptible increase of sensible heat should be produced by friction or hammering except when a very violent force is applied to a small quantity of matter & as such a force can hardly help being attended with a tearing off or displacing of some [of] the particles of those bodies it can hardly help being attended with an alteration of latent heat in the body <which will make the sensible heat produced> very different from what would otherwise be produced by the vis viva communicated by the friction & hammering It seems as if in fact the heat produced was probably greater than what could arise from that cause & consequently that friction & hammering are usually attended with a diminution of latent heat It is possible they may sometimes be attended with an increase of latent heat & consequently that the sensible heat produced will be less than would otherwise take place from the quantity of vis viva communicated*

*The nature of friction & imperfect elasticity deserves to be considerd more accurately

According to Mr Mitchell & Boscovich[8] there are no impenetrable particles of matter butt matter consists only in certain degrees of attraction & repulsion directed towards central points There is the utmost reason to think that this opinion is just but if it is not & it must be admitted that there are solid impenetrable [particles] still there seems sufficient reason to think that those particles do not touch each other but are kept from ever coming in contact by their repulsive forces This being the case it does not readily appear why bodies rubbing on one another should have

On the expansion of bodies by heat & of the different forms which they assume according to their heat

As the particles of which bodies consist are in perpetual vibration it is evident the attraction & repulsion of any two particles on each other must be perpetually varying according to the part of the vibration they are [in] & even their mean attraction & repulsion can hardly be the same as if they were at rest in their mean position so that an increase of heat in a body can hardly fail of being attended with an alteration in the distance or arrangement of its particles so that it <would be very extraordinary if the> is natural to expect that an increase of heat in a body <did not> should cause some alteration in its bulk Why this alteration is always in excess & why the size of the body should

any friction for suppose 2 bodies whose surfaces are full of prominent particles to rub on one another then when any prominent particle of the upper surface falls into a hollow between 2 prominent particles of the lower surface it will require force to draw it out of that hollow or will cause a retardation to the motion of the upper surface but then when the same particle is sinking down into a hollow it will accelerate the upper body & on the whole it is plain that it must as much accelerate that body while sinking into the hollows as it retards it when rising out of them & moreover while some particles are rising out of hollows others will be sinking into them so that when the body is once put in motion there is no reason why it should not continue to do so without meeting with any resistance from friction it is possible however that when one body rests on another it will acquire such a position that the numbers of particles to be drawn out of hollows may exceed that of those tending to fall down into them so that on the whole it may at first require some force to [put] the body into motion The only cause of friction which I can see is that no 2 bodies can rub against each other without either tearing off or displacing some of their particles I imagine however that except in violent frictions few particles are torn off but only displaced

In like manner when 2 bodies impinge against each other or when any body is bent the sole cause of the want of elasticity observed in those cases is the displacing or tearing off of some of the particles of the body

This being premised it seems certain from prop. that whenever any visible vis viva is lost by the rubbing or striking of 2 bodies against each other or by want of elasticity when they are bent those bodies must receive an increase of total heat equivalent to it but as this diminution of visible vis viva cannot take place without displacing some of their particles an alteration of latent heat can hardly fail of taking place so that it would be no wise extraordinary if in some cases it was to be attended with a diminution of sensible heat & it is most probable that the heat which they will receive will be by no means inversely as their capacities for heat as it would otherwise be never or at least scarce ever be diminished thereby I can not pretend to explain but there certainly is nothing <extraordinary in it or what can be urged as an objection to the theory> in it which seems repugnant to the theory[9]

<On the effect of heat in promoting chymical decompositions & combinations>

It also is natural to expect that when the vibrations of the particles are sufficiently great their mutual attractions & repulsions should be so much alterd as to oblige the particles to assume a quite different arrangement so as to totally alter the appearance & properties of the body as is the case in the evaporation & melting & hardening of bodies It may be observed that in general bodies grow more fluid or less hard & their particles acquire a less strong adhesion to each other as their heat increases It may also be observed that most chymical decompositions & combinations are promoted by heat both of which seem no unnatural effects of an increased vibration of the particles

Conclusion

It has been shewn therefore by as strict reasoning as can be expected in subjects not purely mathematical that if heat consists in the vibration of the particles of bodies its effects must be strikingly analogous & as far as our experiments yet go in no case contradictory to the phenomena for first bodies must retain their heat without increase or diminution until alterd either by receiving heat from or communicating it to other bodies or by some other external cause or else by an alteration produced in the arrangement of its particles 2ndly If 2 bodies of different heats are placed in contact one will communicate heat to the other till they both acquire the same sensible heat that is till they shew the same heat by the thermometer 3rd It is reasonable to expect that it should require different quantities of heat to communicate the same sensible heat to different bodies & that the chymical union of different bodies or a change in the nature of any body should be commonly attended with a change of sensible heat 4th It is proved that no heat should be produced by light reflected from or transmitted through a body but only by the light absorbed by it 5th the phenomena of the heating of bodies by friction & hammering nowise disagree with but rather agree with the theory 6th it is shewn to be probable that an increase of heat in a body ought to be attended either by an alteration of its bulk or some other of its properties or both & that a sufficient increase of heat should intirely alter the nature of the body

All these circumstances agree perfectly with the phenomena & I do not know any one phenomenon which seems at all inconsistent with the theory though it is not of that pliable nature as to be easily adapted to any phenomena On the other hand the various hypotheses which have been formed for explaining the phenomena of heat by a fluid seem to show that none of them can be very satisfactory & though it does not seem impossible that a fluid might exist endued with such properties as to produce the effects of heat yet any hypothesis of that kind is of that unprecise kind as not to admit of being reduced to strict reasoning & examining whether it will explain the phenom or whether it will not rather be attended with numberless inconsistencies & absurdities So that though it might be natural for philosophers to adopt such an hypothesis when no better offerd itself yet when S^r I[saac] N[ewton] has proposed a <hypothesis> theory which may be shewn by strict reasoning must produce effects strongly analogous to those observed to take place & which seems no wise inconsistent with any <of the phenomena> there can no longer be any reason for adhering to the former

But to put the matter in a stronger light It seems certain that the action of such rays of light as are absorbed by a body must produce a motion & vibration of its particles among one another <It seems certain also that the chymical mixtures of different substances must commonly either increase or diminish that motion> so that it seems certain that the particles of bodies must have such motion & their motion must produce effects <much> analogous to those actually observed & seems sufficient to account for all Why therefore should we have recourse to the hypothesis of a fluid which nothing proves the existence of <& which for ought we know> when a circumstance which certainly exists in bodies seems fully sufficient to account for the phenomena

<Another strong argument is this> Again It seems certain as was before said that the particles of bodies must have an internal motion & that this motion must produce effects which can hardly fail of manifesting themselves to our senses but no other phenomena occur which can with any probability be attributed to this cause which is another strong argument for supposing that heat must consist in this motion of the particles of bodies

On the heat produced by electricity[10]

When an electric jar is discharged through a wire or other substance it is plain that a certain velocity & consequently a certain quantity of vis viva must be given to the electric fluid in that wire & that vis viva can hardly fail of being finally communicated to the particles of matter in the wire & consequently heating it so that if the vis viva communicated to the electric fluid in the wire is sufficiently great the wire will receive a violent heat but though the vis communicated to the electric fluid should not be sufficient to produce any sensible increase of total heat yet it seems not impossible that the particles of the wire may be displaced thereby in such manner as to cause a great diminution of latent heat & consequently a great increase of sensible heat

It will be proper therefore to endeavour to examine what degree of [vis] viva can be given to the electric fluid by the discharge in doing which I shall argue upon the principle laid down in my paper concerning the cause of electricity in Ph[ilosophical] Tr[ansactions].[11]

Let ACB & ADE be 2 wires communicating at A with [the] positive side of an electric jar while the negative side communicates with the ground & the 2 wires touch each other at B They will become overcharged & will repel each other with a force equal to the repulsion of the redundant fluid in ACB on the redundant fluid in ADB Take now the point E such that the

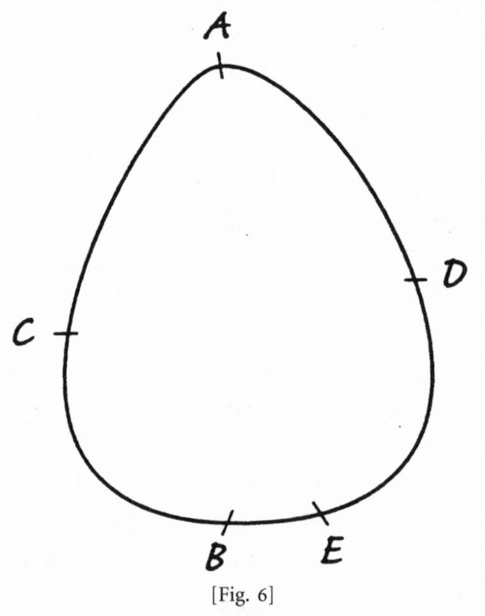

[Fig. 6]

dist. to which repulse extends = d
red. flu. in D° = col. length ——— r
total flu. in D° ——————— t
fluid discharged ——————— f
repuls = weight col. whole length = g
therefore fluid t is urged by its weight $\times$ g/r through space $= f \times \frac{d}{t}$ therefore

$$\text{vis viva} = \frac{tg}{r} \times f \times \frac{d}{t} = \frac{gfd}{r}$$

$$\text{force} = \frac{r}{f}$$

$$\text{space} = \frac{gd}{\phi}$$

repulsion of redundant fluid in ACB on the redundant fluid in ADE shall be small in comparison of its repulsion on that in BE so that the repulsion of the 2 wires may be considerd as equal to the repulsion of the redundant fluid in ACB on the redundant fluid in BE Let the repulsion of the 2 wires be equal to the weight of a piece of the wire BE whose length = r & let the redundant fluid on the positive side of the jar the redundant fluid in BE & the whole fluid in BE be equal in weight to pieces of the same wire whose lengths are g f & ϕ

respectively Let now the wire ADB be removed & made to communicate by its end A with the negative side of the jar & let its end B be suddenly brought in contact with the same end B of the other wire so as to discharge the jar Now the redundant fluid on the positive side of the jar is equal to the whole fluid containd in a piece [of] the wire whose length $= d \times \frac{g}{\phi}$ so that the fluid in the wire will move in consequence of the discharge through the space $\frac{gd}{\phi}$ Moreover each particle of fluid in BE is repelled from B with the same force that the particles of the redundant fluid were repelled in the former state of the wire & are therefore repelled with a force which is to their weight as r is to f; & the whole force with which they are impelled is equal to the weight of a piece of wire whose length $= r \times \frac{\phi}{f}$ so that the vis viva given to the fluid in BE is equal to $\frac{r\phi}{f} \times \frac{gd}{\phi} = r \times \frac{dg}{f}$

NB $\frac{dg}{f}$ is the length of a piece of wire of the same thickness as BE which if electrified with the same force as the jar would receive as much redundant fluid as is collected on the positive side of the jar It must be observed also that the vis viva given to the electric fluid in the whole length of wire ADB is no more than what is given to that in BE for as the fluid in the part ADE is impelled only by that in BE it can receive no vis viva without taking away as much from that in BE

It should seem from hence that the vis viva given to the electric fluid by the discharge is by no means sufficient to account for the heat produced except by displacing the particles & thereby causing a diminution of latent heat

The velocity given to the electric fluid in BE is that which a body acquires by its gravity in falling from the height $\frac{dg}{f} \times \frac{r}{\phi}$ so that if the weight of the electric fluid naturally containd in a body is excessively small in proportion to the whole weight of the body as is not unlikely to be the case the velocity given to the electric fluid will be very great but not otherwise so that it is uncertain whether the velocity given to the electric fluid is great or not

It is commonly supposed that the velocity of the electric fluid is very great but there is nothing which shows whether it is or not it has been found indeed that if a jar is discharged through a very long wire interrupted in the middle & each end, there is no sensible interval of time between the appearance of the spark at the middle & at the ends which only shews that the pulse as we may call it is propagated with very great velocity through the wire but this does not at all depend on the velocity with which the fluid itself moves any more than the velocity of sound depends on the velocity with which the particles of air vibrate

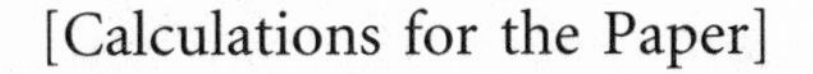

[Calculations for the Paper]

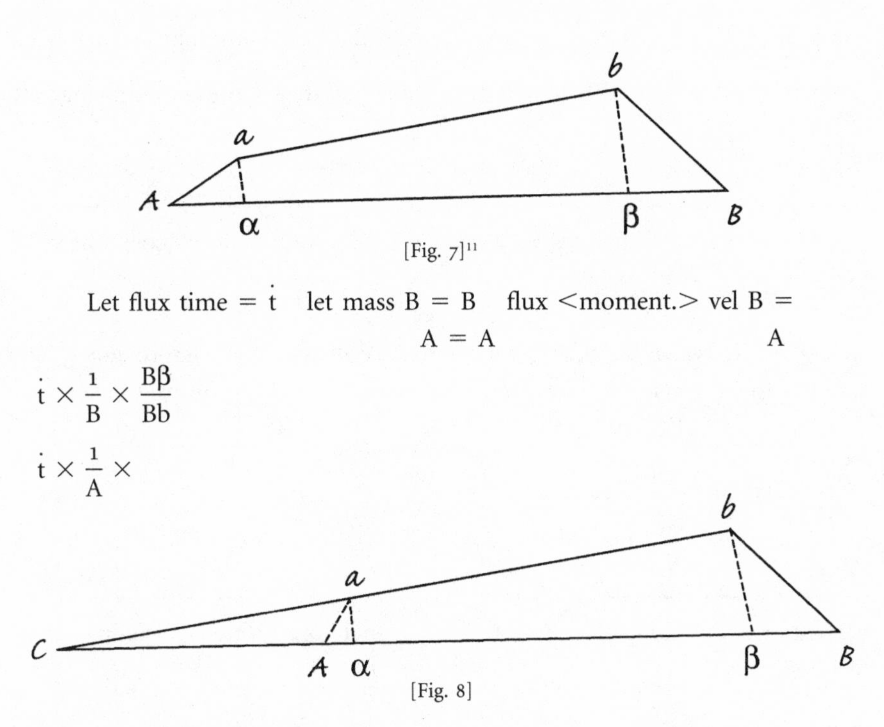

[Fig. 7][11]

Let flux time = $\dot{t}$ let mass B = B flux <moment.> vel B =
A = A A

$$\dot{t} \times \frac{1}{B} \times \frac{B\beta}{Bb}$$

$$\dot{t} \times \frac{1}{A} \times$$

[Fig. 8]

Increase moment. of body drawn by given force in given time is the same whatever is the weight

Increase mom. B is same as acquired in moving through Bβ & incr. mom. A is same as in moving through Aα & therefore incr. of sum of their momenta is the same which either of them would acquire in moving through BA − ba

---

Let $\dot{B}$ = increase moment. B

$\overline{bd}^{\cdot}$ mom. B owing to increase action of d

$\overline{\beta d}^{\cdot}$ increase mom. which B would acquire by D° in moving through Bβ

$\overline{BD}^{\cdot}$ = incr. mom. which B or D would acq. in moving through BD-bd then

$\overline{bd}^{\cdot} - \overline{\beta d}^{\cdot} = 0$ & conseq

$\overline{bd}^{\cdot} + \overline{ba}^{\cdot} - \overline{\beta d}^{\cdot} - \overline{\beta a}^{\cdot} - \&^{ct} = 0$

$\overline{db}^{\cdot} + \overline{da}^{\cdot} - \overline{\delta b}^{\cdot} - \overline{\delta a}^{\cdot} - \&^{ct} = 0$

$\overline{ab}^{\cdot} + \overline{ad}^{\cdot} - \overline{\alpha b}^{\cdot} - \overline{\alpha d}^{\cdot} - \&^{ct} = 0$

but $\overline{bd}^{\cdot} + \overline{ba}^{\cdot} = \dot{B}$

$\overline{\beta d}^{\cdot} + \overline{\delta b}^{\cdot} = \overline{BD}^{\cdot}$

therefore $\dot{B} + \dot{D} + \dot{A} - \overline{BD}^{\cdot} - \overline{BA}^{\cdot} - \overline{DA}^{\cdot} = 0$

---

let vel. light = v inc. per " weight falling in 1" = w then impulse of light would in 1" give vel. $v \times \frac{w}{a}$ to weight a & if the light communicates its whole momentum to weight a its momentum will equal that of weight a moving with vel $v\sqrt{\frac{w}{a}}$

---

light from 1½ sq. foot does in 1" give vel 1 inc. per " to 10 gra therefore $\frac{vw}{10} = 1$ or $w = \frac{10}{v}$

weight col. water whose base = 1½ sq. foot & alt. = 1 inc = 55000 gra & light should in 1" give to it same moment. as if it moved with vel. $v\sqrt{\frac{10}{v \times 55000}} = \sqrt{\frac{v}{5500}}$ but v = 12000.000.000 & $\sqrt{\frac{v}{5500}} =$

$\sqrt{\frac{24000.000}{11}} = \sqrt{2200.000} = 1500$

---

vel acquired by falling 16 feet = 32 × 12 therefore mom light falling on 1½ sq. feet in 1" = that of weight $a \times \frac{v^2w}{a \times (32 \times 12)^2} = \frac{v^2w}{(32 \times 12)^2}$ but vw = 10

& therefore $\frac{v^2w}{(32 \times 12)^2} = \frac{120.000.000.000}{(32 \times 12)^2} = \frac{10.000.000.000}{32 \times 32 \times 12} = \frac{10.000.000.000}{12.288}$
= 800.000 × 16 feet which if we suppose a man of 12 stone or 12 × 14 × 7000 = 24 × 49000 = 1200.000 gra. can raise his own weight 2 feet per " is equal to the labour of $\frac{800.000 \times 8}{1200.000} = \frac{16}{3}$ such men

[Changes in the Paper]

Addit & alter in foul copy

P. 8 x Note When 2 substances which have a chymical affinity unite together it seems likely that heat & not cold should commonly be the consequence for unless the attracting particles approach nearer together or the repelling particles recede further so as that the value of S shall be increased one does not easily see why the 2 bodies should mix but if the value of S is increased, a diminution of latent heat and an equal increase of active heat must take place Accordingly except where one of the bodies is changed by the mixture from a solid to a fluid form or from either of those forms to that of an elastic fluid I do not recollect a single instance of cold being produced by any chymical mixture <Perhaps too a reason may be assigned why cold should always be produced by a change from a solid to a fluid form for as the quantity of latent heat in a body must in all probab. be very different when it is in a fluid form from what it is when in a solid form the quantity of total For the same reason it may naturally be expected that the change of a body from a solid to a fluid form or from either of those forms to that of an elastic fluid should be attended with a great alteration of sensible heat & this as well as the change of heat caused by chymical mixtures may proceed either from its requiring more active heat to give the same sens. heat to a body in one form than in the other or from the quantity of latent heat being very different in those 2 forms It is more difficult to assign a reason why the change of form just now mentiond should always be attended with an increase of sensible cold & never of heat But as cold is frequently produced by chymical mixtures in which one or both of the bodies undergo this change one may be inclined to think that it is owing to a body requiring more active heat in order to produce a give[n]> But in mixtures in which this change of form ensues it is well known that sensible cold is frequently produced but if this increase of sensible cold proceeds from an increase of latent heat one does not well see why the mixture should take place which might incline one to think that the cold produced by this change of form was owing to its requiring a greater quantity of total heat to produce a given sensible heat to a body when it is in a fluid than in a solid form & still more when it is in the form of an elastic fluid

As the reasoning in this note is too hypothetical & is not material to the main purpose of this paper I chose to put it in this form rather than to insert it in the text

Addit to P. 16) The circumstance that rays of light <are emitted only from very hot bodies> & heat are emitted more plentifully as the heat increases agrees very well with the theory for this effect cannot be produced without these particles being removed from that situation in which they are retaind in their places by the attractions & repulsions of the other particles of the body & coming into that position in which they receive the violent repulsion necessary to emit them with their proper velocity Now while the particles of the body are at rest there seems no cause which should produce that effect but when they are in motion it is not extraordinary that the particles of light should sometimes come into the position necessary to give them this repulsion & that they should do so the more frequently as that motion increases for the particles cannot be emitted without being removed by some cause from their natural situation in which they are kept in their places by the attractions & repulsions of the other particles of the body & being brought into some other position in which [they] receive that violent repulsion necessary to give them their proper velocity

---

P. 35) line 7) Moreover each particle of fluid in BE is repelled from B with the same force that the particles of redundant fluid were repelled in the former position of the wire & are therefore repelled with a force which is to their weight as r to f & the whole force with which they are repelled is equal to the weight of a piece of the wire whose length $= \phi \times \frac{r}{f}$ Therefore the velocity which this fluid acquires by the discharge must be that which a body acquires by gravity in falling from the height $\frac{gd}{\phi} \times \frac{r}{f}$ & the vis viva which the whole electric fluid in BE acquires by the discharge is the same which a piece of the wire whose length $= \phi$ acquires by falling from the height $\frac{gdr}{f\phi}$ or which a piece whose length is gd/f acquires by falling from the height r

---

P. 36 par. 3 The force with which the electric fluid in BE is impelled was before said to be equal to the weight of a piece of the wire whose length $= r\phi/f$ & therefore is most likely very great as the quantity of redundant fluid in an electrified body is most likely very small in comparison of the whole quantity naturally containd in it The velocity given to this fluid was before said to be that acquired in falling from the height $\frac{dg}{f} \times \frac{r}{\phi}$ & therefore if the weig[ht]

But this does not shew any thing as to the real velocity given to the electric fluid for we are not to suppose that the individual electric fluid which issues from the positive side of the jar passes through the whole length of the wire & enters in the negative side <of the jar> on the contrary the space through which the electric fluid passes is in all probability not great but it serves to push forward the electric fluid before it & thereby to propagate the motion through the wire just as the motion of the particles of air propagate sound & the swiftness with which the motion is propagated through the wire does not at all depend on the velocity of the elastic fluid anymore than the velocity of sound depends on that with which the particles of air vibrate [This statement is to be inserted after "at the ends," p. 170, l.1.]

[REVISED COPY OF THE MANUSCRIPT]

[Heat]

Preliminary propositions

Def. 1) The effect which a body in motion can produce in mechanical purposes is as the weight of the body multiplied by the square of its velocity, or as its weight multiplied by the height from which it must fall by its gravity to acquire that velocity. This force I shall call its vis viva; & for shortness, if the weight of the body is g grains, & its velocity is such as it would acquire by gravity in falling f. feet, I shall call its vis viva g grains x f. feet & so on

Def 2) In computing the vis viva of a system of bodies, the vis viva of each body must be taken separately, by multiplying its weight into the square of its velocity, without regard to the direction of that velocity; & the sum is the vis viva of the whole system

Pr 1) The increase of vis viva which a body already in motion acquires in moving through a given space by a given force is equal to the whole vis viva which it would acquire in moving from rest through the same space

Prop. 2) If 2 bodies attract, & in consequence move towards each other, the smaller body will acquire thereby a greater vis viva than the greater, in the inverse ratio of their weight

Prop. 3) If there are 2 bodies A & B, & one of them as A is fixed, & the other B moves towards it till it comes within a certain distance of it, the vis viva which it will acquire is just the same that A would acquire in moving to within the same distance of B, if B was fixed & A at liberty

Pr. 4) Let there be a body whose particles are in motion amongst each other, but in such manner that their center of gravity is at rest; & let the vis viva of the body be called S. Let now a new velocity V be given to each particle in a given direction; by which means the motion of the particles in respect of each other will not be alterd; but the center of gravity of the body will be made to move with the velocity V; & let the weight of the whole body multiplied by the square of V be called T. Then will the whole vis viva of the body be S + T

For let B (Fig. 1) be any particle of this body. Suppose

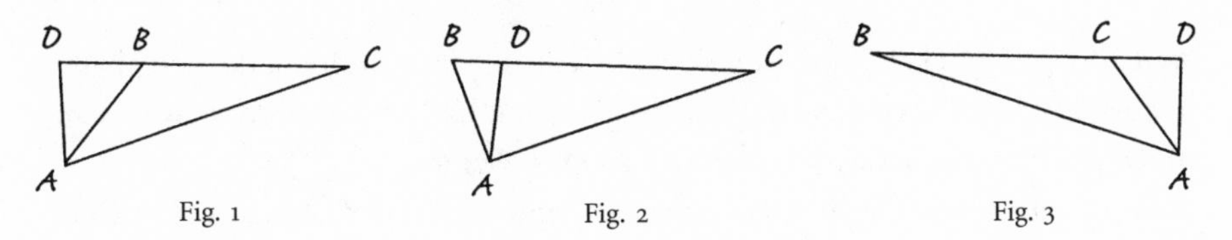

Fig. 1 Fig. 2 Fig. 3

that before it received its new velocity it was moving with the velocity AB in the direction AB; & let the new velocity V given to it be equal to BC, & in the direction BC. Therefore after that, it will move in the direction AC with the velocity AC; & therefore if its weight is called w, its vis viva will be $w \times AC^2$; which is equal to $w \times AB^2 + w \times BC^2 + 2w \times BD \times BC$; AD being drawn perpendicular to BC, & observing, that if D is on the same side of B that C is, as in fig. 2 & 3, then BD must be considerd as negative. But the sum of all the quantities $w \times AB^2 = S$; & the sum of all the quantities $w \times BC^2 = T$; & the sum of all the quantities $2w \times BD \times BC = 2^{ce}$ the weight of the body multiplied by the product of the velocity which the center of gravity had in the direction BC before the application of the new velocity, & of the velocity V; & which quantity by the supposition is nothing. Consequently the whole vis viva of the body after the application of this new velocity is $S + T$

Definit) Let there be a body whose particles are in motion. Let G be its center of gravity; & GN any line drawn through it, which we will call its axis. Let B be any particle, & BM a

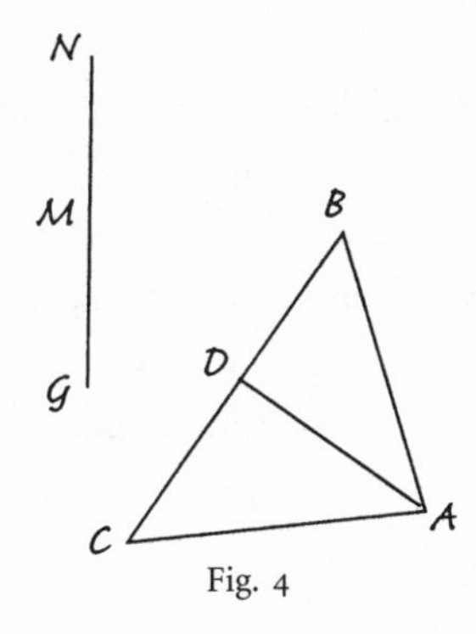

Fig. 4

perpendicular from thence to the axis; & let BD be the rotatory velocity of that particle; that is let it be equal to that part of its velocity which is performed in the direction BD, perpendicular to BM & in a plane perpendicular to the axis, & which is the direction in which it must move if it revolved round the axis. Let now the weight of B be multiplied by the product of BM & BD; & let the same thing be done by each other particle; observing that if the direction of its motion is such as to make it revolve in the contrary direction from B, this product must be considered as negative. Then if the sum of all these products is nothing, the body is said to have no rotatory vis viva on the axis GN & if it has no rotatory vis viva on any other axis it is said to have no rotatory vis viva whatever

Pr. 5) Let the particles of this body be in motion, but in such manner to give it no rotatory vis viva, & in such manner that its center of gravity is at rest; & let its vis viva be called S. Let now a rotatory velocity be given to each particle proportional to its distance from the axis; by which means the body will acquire a rotatory vis viva without altering the motions of its particles in respect of each other; & let T be equal to what would be the vis viva of the body if its particles had only this rotatory motion. Then will the whole vis viva of the body be S + T

For take any particle B; let its weight be w; & suppose that before the new motion given to it, it was moving in the direction AB with the velocity AB; consequently if AD is drawn perpendicular to BC, BD was its rotatory velocity. Let now the new rotatory velocity given to it be equal to BC, which by the supposition is proportional to BM. Then its vis viva before the application of this new motion was w × AB[^{2}]; & after it, is w × AC^2; which is equal to w × AB^2 + w × BC^2—2w × BD × BC but the sum of all the quantities w × AB^2 = S; & the sum of all the quantities w × BC^2 = T; & the sum of the quantities 2w × BD × BC, or 2w × BD x BM, is by the supposition equal to nothing; & therefore the whole vis viva of the body is S + T

Coroll.) In both these propositions the quantity S may properly be called the invisible vis viva of the body, as it arises only from the motion of the invisible particles of the body among each other; & in like manner the quantity T may be called the visible vis viva. Now it appears from these 2 propositions, that whether the body has a progressive or rotatory vis viva, its total vis viva is equal to the sum of its visible & invisible vires vivae; & the case is evidently the same though the body has a progressive & rotatory vis viva both together. It follows therefore that if the visible vis viva is diminished without altering the total vis viva, the invisible vis viva must be as much increased

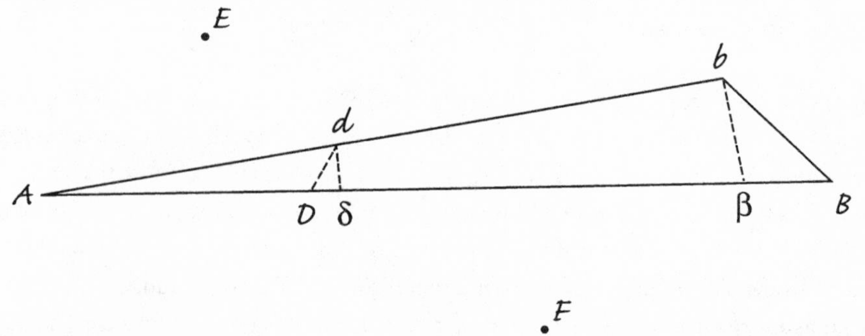

Fig. 5

Pr. 6 Let the particles B D E & F fig. [5] attract or repel each other in such manner that the force with which any 2 particles attract or repel each other is every where the same at the same distance however different at different distances. Let B, D, E & F be the present positions of those particles; & suppose that at the end of a given very short time B & D are removed to b & d. Let BD & bd continued meet in A; & let this time be so small that the angle bAB shall be very small; & with the center A draw the arches $d\delta$ & $b\beta$. Then $B\beta + D\delta$ is the space by which B & D approach each other; observing that if β is further from D than B is, so that the motion of B is such as to make it recede from D, $B\beta$ must be considered as negative; & the like must be observed with regard to $D\delta$. Let now $\dot{B}$ be the increase of vis viva which B receives in moving over Bb by the united action of the other particles. Let $\overline{bD}$ be the increase of vis viva[13] which it would receive in moving over the same space by the action of D alone, supposing it to be confined to move in the direction Bb, but without being acted on by E or F; & let $\overline{\beta D}$ be the increase of vis viva which B would receive in moving through the space $B\beta$ by the action of D supposing it to set out with any velocity whatever;* observing that if B & D repel each other, their mutual action must be considerd as negative; & therefore that if $B\beta$ & the mutual action of B & D are both positive or both negative, $\overline{bD}$ & $\overline{\beta D}$ must be considered as positive, but otherwise negative, as in the 1st case the increase of vis viva expressed by the quantities $\overline{bD}$ & $\overline{\beta D}$ will be positive and in the latter negative. Moreover let $\overline{BD}$ be the increase of vis viva which either B or D would acquire by falling through $BD - \beta\delta$ by their mutual action;† & let the increments of vis viva of the other particles be represented in the same manner by their respective letters

Now the increase of vis viva which B receives in consequence of the action of D while it describes the line Bb, is equal to that which it would receive by the same cause in describing the line $B\beta$; & therefore $\overline{bD} = \overline{\beta D}$ or $\overline{bD} - \overline{\beta D} = 0$ & therefore

$$\overline{bD} + \overline{bE} + \overline{bF} - \overline{\beta D} - \overline{\beta E} - \overline{\beta F} = 0$$
$$\overline{dB} + \overline{dE} + \overline{dF} - \overline{\delta B} - \overline{\delta E} - \overline{\delta F} = 0$$
$$\overline{eB} + \overline{eD} + \overline{eF} - \overline{\varepsilon B} - \overline{\varepsilon D} - \overline{\varepsilon F} = 0$$
$$\overline{fB} + \overline{fD} + \overline{fE} - \overline{\phi B} - \overline{\phi D} - \overline{\phi E} = 0$$

*Vide Prop. 1.

†Vide Prop. 3

But $\overline{bD} + \overline{bE} + \overline{bF} = \dot{B}$ & answers to an increase or decrease of vis viva according as it comes out positive or negative; & therefore $\dot{B} + \dot{D} + \dot{E} + \dot{F} - \overline{\overline{\beta D} - \overline{\delta B}} - \overline{\overline{\beta E} - \overline{\varepsilon B}} - \overline{\overline{\beta F} - \overline{\phi B}} - \overline{\overline{\delta E} - \overline{\varepsilon D}} - \overline{\overline{\delta F} - \overline{\phi D}} - \overline{\overline{\varepsilon F} - \overline{\phi E}} = 0$. But $\overline{\beta D} + \overline{\delta B} = \overline{BD}$, & answers to an increase or decrease of vis viva according as BD comes out positive or negative; & therefore $\dot{B} + \dot{D} + \dot{E} + \dot{F} - \overline{BD} - \overline{BE} - \overline{BF} - \overline{DE} - \overline{DF} - \overline{EF} = 0$. But $\overline{BD}$ is half the sum of the increments of vis viva which B would acquire in moving through BD − bd by their mutual action & which D would acquire by the same means; & therefore the sum of the increments of vis viva which the particles B, D, E, & F receive during this short time by their mutual action, minus ½ the sum of the increments of vis viva which each particle would receive by the action of each other particle in moving through the distance by which they approach each other, remains unalterd

Coroll) Though for simplicity I have considerd the action of only 4 particles on each other, yet it is evident that the case would be exactly the same though their number was ever so great.

Therefore let there be a body consisting of any number of particles attracting or repelling each other as above. Let the increase of vis viva which each particle would acquire in falling by the attraction or repulsion of each other particle from an infinite distance to its actual distance from that particle be computed separately;* & let ½ the sum of these increments of vis viva be called S. Let these particles be supposed in motion; & let the sum of their actual vires vivae be called s. Then s − S cannot be alterd by their motions among each other. It must be observed however that if there is any such thing as solid impenetrable particles of matter,† & that these particles can ever impinge against each other, those particles must suffer a loss of vis viva, & consequently the value of s − S would be continually diminishing. But it seems impossible that any such blow should ever take place

*It is possible that the particles may constantly attract or constantly repel each other, or that they may alternately attract & repel as the distance varies. Whichever of these is the case, if on the whole the particle suffers a decrease of vis viva by the fall this quantity must be considerd as negative.

My reason for directing increments of vis viva to be computed is that if on the whole the particle is repelled it could not make this fall unless it set out with some velocity. But as was said before this increment of vis viva is just the same whatever velocity it sets out with

†See further on under the head of heat produced by friction

Hypothesis

1) Heat I suppose consists in the internal motion of the particles of which bodies are composed; but though these particles are in constant motion, yet they are so far retaind in their place by their mutual attractions & repulsions, that while the nature of the body remains unchanged, the greatest distance to which they can be removed by these vibrations from their original situation, must be very small

2) By the sensible heat of a body I mean its heat as shewn by a thermometer

By the quantity of latent heat in a body I mean the value of $-S$

By the quantity of active heat I mean the value of s; & consequently $s - S$ is the quantity of total heat

3) It was before said that the value of $s - S$ or the quantity of total heat must remain constantly the same until alterd by some external cause, & that it is not alterd by the motion of the particles amongst each other; but strictly speaking the values of s & S or the quantities of active & latent heat must be continually varying, & can never remain exactly the same even for an instant. Yet as the number of vibrating particles even in the smallest body must be inconceivably great, & as the vis viva of one must be increasing while another is diminishing, we may safely conclude that neither of them can sensibly alter, except from some external cause, unless the general size or arrangement of the particles of the body are alterd. Then indeed as the value of $-S$ or quantity of latent heat can hardly escape being alterd, the quantity of active heat must be alterd as much

On the communication of heat

4) If 2 bodies A & B precisely of the same nature, but unequally hot, be put in contact, it seems clear that the particles of the hottest body will communicate motion to those of the colder, till the particles of both come to vibrate with the same velocity, that is till both bodies have the same quantity of active heat in proportion to their bulk; provided neither are in contact with any 3rd body from which they can receive, or to which they can communicate motion

5) But if the bodies are of different natures, it seems likely that they should never acquire the same quantity of active heat in proportion to their bulk, though they are kept ever so long in contact.* But still though A should not

*For example if the vibrating particles of B are less than those of A, it is likely that B will receive more active heat in proportion to its size than A, as seems probable from the consideration of Prop 2. It is likely also that the proportional

acquire the same quantity of active heat in proportion to its bulk as B; yet if a 3rd body C is put in contact with A & a 4th body D precisely of the same nature as C is placed in contact with B, it seems reasonable to suppose that C will acquire the same quantity of active heat in proportion to its bulk, as D so that A & B will still shew the same degree of heat by the thermometer

For let us suppose that when 2 bodies of equal weights remain long enough in contact, the proportional quantity of active heat in them will be as some function, either of the size of their particles or of any other quality in them. Let this function answering to the 3 bodies A B & D be α β & δ; & for greater simplicity let all the 4 bodies A B C & D be of the same weight. Then as C is of the same nature as D the quantity of active heat in it will be to that in A as $\delta:\alpha$; & the active heat in B will be to that in A as β to α; & that in D will be to that in B as δ to β; & therefore the active heat in D will be to that in A as δ to α, & therefore will be equal to that in C

6) From what has been said it appears first that when 2 bodies unequally heated & of ever so different natures are brought in contact, the hotter body must communicate heat to the other till they both acquire the same sensible heat, provided neither of them are in contact with other bodies which can communicate to or carry off heat from it. 2ndly the quantity of total heat which one body parts with must be equal to that which the other acquires. 3rdly some bodies may require a greater addition of total heat to produce in them a given increase of sensible heat than others; & this from 2 causes; first that some bodies may require a greater addition of active heat than others in order to produce the same increase of sensible heat, as was before said in art. 5 & 2ndly because in all bodies an alteration of sensible heat can hardly help being attended with an alteration of the quantity of latent heat. For as the bulk of all or at least almost all bodies is increased by heat, the distance of their particles must be alterd; which can hardly fail of being attended by an alteration of the value of S, that is of their latent heat; & that alteration can hardly fail of being greater in some bodies than others

On the heat & cold produced by chymical mixtures &
by a change from a solid to a fluid form

7) It seems a natural consequence of this theory that the mixture of two substances which have a chymical affinity should commonly be attended by an

quantity of active heat in the 2 bodies may be affected by the time in which their particles vibrate alteration of sensible heat; for as the arrangement of the particles must be alterd thereby, the quantity of latent heat can hardly fail of being alterd; & moreover it is very possible that the quantity of active heat necessary to produce a given sensible heat, may also be alterd; both of which causes will produce an alteration in the quantity of total heat necessary to produce a given sensible heat; & consequently as the quantity of total heat remains unchanged, the sensible heat must be alterd

8) For both these reasons also it seems a necessary consequence of the theory that the change of a body from a solid to a fluid form or from either of those forms to that of an elastic fluid should commonly be attended with an alteration of sensible heat*

On heat from the impulse of light

9) If a body A is at rest & another body B moves towards it, & when it comes near it is repelled & either reflected straight back or turned out of its course; then if A is not much bigger than B, B will lose a great part of its vis viva, & A will acquire as much. But if A is excessively bigger than B, B will lose an excessively small part of its viva, & will communicate excessively little to A; & in all cases A will acquire just as much vis viva as B loses

There can be no doubt that light consists of excessively small particles emitted with excessive velocity from the luminous body; & it has been sufficiently proved

*When 2 substances which have a chymical affinity unite, it seems likely that heat & not cold should commonly ensue; for unless the attracting particles approach nearer together or the repelling particles recede further, so as to increase the value of S, one does not easily see why the 2 bodies should mix. But if S is increased, the quantity of active heat must be equally increased; & consequently the sensible heat will in all probability be increased. This agrees with observation; for except where one of the bodies is changed by the mixture from a solid to a fluid form, or from either of those forms to that of an elastic fluid, I do not know a single instance of cold being produced by any chymical mixture. But in mixtures in which this change of form takes place, it is well known that cold is frequently produced. But if this increase of sensible cold proceeds from an increase of latent heat, one does not well see as was before said why the mixture should take place; which might incline one to think that the cold which always attends this change of form proceeds from the latter of the abovementiond causes, or to more active heat being necessary to produce a given sensible heat when the body is in a fluid than a solid form

As the reasoning in this note is too hypothetical & is not material to the main purport of the paper I chose to put it in this form rather than insert it in the text.

that they are not reflected by impinging against any solid particles or even by the repulsion of a few small particles only, but by the joint repulsion of a quantity of matter infinitely greater in quantity than the particles of light themselves, so that they can lose no sensible part of their vis viva thereby, nor can communicate any sensible vis viva to the body; & the same thing takes place when light passes through a transparent body & is refracted. But when light enters into a body & is stopt; then the particles of light will be continually reflected backwards & forwards till they at last come to vibrate with no greater velocity than the particles of the body itself; so that their vis viva will be equally distributed between the body & them

It follows therefore that a body can receive no heat from light reflected from it or refracted through it, but only from that part of the light which is stopt in it & that the total heat produced thereby is equal to the vis viva of the particles stopt in it

10) It was said that bodies are not heated by rays transmitted through them; but yet it is well known that a plate of glass is much more heated by the fire, & I believe by the ☉, than a plate of polished metal, though the metal plate absorbs more light than the glass; which at first sight seems contradictory. But the reason of this is satisfactorily explained by the observations of Mr Scheele; as he has proved that hot bodies emit not only rays of light, but also other particles, which though not capable of exciting the sensation of light in our eyes are yet able to produce heat, & which may therefore be called rays of heat; he has shewn too that these rays of heat are reflected by polished metals, but are neither reflected nor transmitted by glass. It therefore is not extraordinary that the glass should be heated more than the metal as it is heated by the rays of heat

The ☉^s^ rays seem to contain a less proportion of rays of heat than those of common fire; & Mr Saussure has found that bodies emit rays of heat though not near hot enough to emit rays of light, & even though not hotter than boiling water; so that it should seem that the more intensely a body is heated the less proportion of rays of heat it emits

11) It seems a necessary consequence of this theory that all bodies exposed to the ☉^s^ rays ought to receive an equal quantity of total heat from it, provided they are equally dark colourd, except so far as depends on some of them absorbing more of the rays of heat than others; but the sensible heat which they will acquire depends also on their capacities for heat, & the ease with which they transmit heat to the air & other bodies in contact with them, that is to the swiftness with which they lose the heat communicated to them

Exper. should be made to examine this

12) In order to find the vis viva of the ☉ˢ light; let the weight of the light which falls in 1" on 1½ sq. feet of surface be called w; & let the velocity of light be v inches in a ". A body will in falling by gravity through 16 feet acquire a velocity of 32 × 12 inches per "; & therefore the vis viva of the light which falls during 1" on 1½ sq. feet $= w \times 16 \times \left(\frac{v}{32 \times 12}\right)^2$ feet. Let now all this light be received on a plate exposed perpendicularly to it whose weight is p; then if this plate reflects all the light, it will by its impulse receive in 1" a velocity of $2v \times \frac{w}{p}$ inches per "; but if it absorbs all the light it will receive only ½ that velocity. Now according to an experiment of Mʳ Michell[17] related by Dʳ Priestley,* it seemed that the impulse of the light falling from the ☉ on a surface of 1½ sq. feet was sufficient to give in 1" of time a velocity of 1 inch per " to a plate weighing 10 gra.; & therefore if we suppose that the plate reflected all the light $\frac{2wv}{10}$ gra must equal 1 or wv must = 5 gra; & therefore according to this experiment the vis viva of the light which falls from the ☉ in 1" on 1½ sq. feet of surface $= 5 \text{ gra.} \times \frac{v}{64 \times 12 \times 12}$ feet, which as v = 12.000.000.000, is equal to 6500.000 gra. × 1 foot & if we suppose that the plate absorbed all the light the vis viva is 2ᶜᵉ as great

If a horse working in a mill can raise 100 £ at the rate of 3 miles an hour or 4½ [feet] a " the labour of a horse in 1" is sufficient to produce a vis viva of 3150.000 gra. into 1 foot; so that the vis viva of the ☉ˢ light falling on 1½ sq. feet is equal to the labour of more than 2 horses.†

Exper. to determine the vis viva necessary to give a given increase of sensible heat to a given body by alternately exposing a thermometer to the ☉ & shading it

It must be observed that as Mʳ Michells experiment was tried under a glass cover it shewed the impulse of the rays of light only & not of those of heat; so that the total vis viva of the ☉ˢ rays must be greater than here computed

*Priestleys optics, P. 387

†If it was possible to make a wheel with float boards like those of a water wheel which should move with ½ the velocity of light without suffering any resistance from friction & the resistance of the air, & as much of the ☉ˢ light as falls on a surface of 1½ sq. feet was thrown on one side of this wheel, it would actually do more work for any mechanical purpose than 2 horses

It is uncertain with what velocity the rays of heat are emitted <If they are emitted with less velocity than those of light it will [make] the vis viva of the ⊙s rays less than according to the foregoing computation>

On the emission of light

It seems most likely that when light is emitted from a body, the particles receive their velocity from the repulsion of a quantity of matter vastly superior in weight to the particles themselves; & if this is the case, no perceptible heat will be produced by the process. But if the quantity of matter by which the particles are repeld is not vastly superior to them in weight, some heat will be generated. It seems impossible to determine by experiment whether any is generated or not

The circumstance that rays of light & heat are emitted more plentifully as the heat increases, agrees very well with the theory. For the particles can not be emitted without being by some cause removed from their natural situation in which they are kept in their places by the attraction & repulsion of the other particles of the body, & being brought into a position in which they can receive that violent repulsion necessary to give them their proper velocity. Now while the particles of the body are at rest, there seems no cause which should produce this effect; but when they are in motion it seems not extraordinary that the particles of light should sometimes come into that position necessary to give them this repulsion, & that they should do so the more frequently as the swiftness of the motion increases

On the heat produced by friction & hammering

Whenever one body rubs against another or whenever 2 unelastic bodies strike against each other it is natural to suppose that the vis viva lost thereby will be spent in communicating motion to the particles of the bodies; or in other words that the invisible vis viva will be as much increased as the visible vis viva is diminished. Though on the other hand if the velocity with which the bodies rub or impinge against each other is less than that with which their particles vibrate, one does not well see why that rubbing or impinging should increase the vibrating velocity of the particles. But whichever of these is the case it seems likely from what was said of the quantity of vis viva requisite to produce a given alteration of sensible heat in bodies, that no perceptible increase of heat should be produced by friction or hammering except when a very violent force is applied to a small quantity of matter; & as such a force can hardly help being attended with a tearing off or displacing of some of the particles of the bodies, the quantity of latent heat will most likely be alterd; the probable consequence of which is that the heat which bodies receive from friction, will in many cases be greater than what ought to proceed from the vis viva communicated thereby to their particles; & that the sensible heat which different bodies will receive from the same degree of friction, will by no means be inversely as their capacities for heat, as would otherwise be the case*

*According to Father Boscovich & Mr Michell matter does not consist of solid impenetrable particles as commonly supposed, but only of certain degrees of attraction & repulsion directed towards central points. They also suppose that the action of 2 of these central points on each other alternately varies from repulsion to attraction numberless times as the distance increases. There is the utmost reason to think that both these suppositions are true; & they serve to account for many phenomena of nature which would otherwise be inexplicable. But even if it is otherwise, & if it must be admitted that there are solid impenetrable particles, still there seems sufficient reason to think that those particles do not touch each other, but are kept from ever coming in contact by their repulsive force

This being the case, it does not readily appear why bodies rubbing on each other should have any friction; for suppose 2 bodies whose surfaces are full of prominent particles to rub on each other; then when any prominent particle of the upper surface falls into a hollow between 2 prominent particles of the lower surface, it will require some force to draw it out, & will cause a retardation in the motion of the upper surface; but then when the same particle is sinking down into a hollow, it will accelerate the upper body, & on the whole it is plain that it must as much accelerate that body while sinking into the hollows, as it retards it while rising out of them; & moreover while some particles are rising out of hollows others will be sinking into them; so that when the body is once put in motion no reason appears why it should not continue to do so without meeting with any resistance from friction; though there is an evident reason why force may be required to first put it in motion, as it may have found a situation in which more particles are to be drawn out of hollows than tend to fall into them

The only cause of friction which I can see, is that no bodies can rub against each other without either tearing off or displacing some of their particles. I imagine however that except in violent frictions few particles are torn off, but only displaced

In like manner when 2 bodies strike each other or when a body is bent the sole cause of the want of elasticity observed is the displacing or tearing off [of] some of the particles of the bodies

This being premised it seems certain from Prop. 6 & the coroll. to Prop. 5 that whenever any visible vis viva is lost by the rubbing or striking of bodies against each other, or by want of elasticity when they are bent, those bodies must receive an augmentation of total heat equivalent thereto. But as this diminution of visible vis

On the heat produced by electricity

When an electric jar is discharged through a wire or other substance, it is plain that a certain velocity, & consequently a certain vis viva must be given to the electric fluid in that wire; & that vis viva can hardly fail of being finally communicated to the particles of matter in the wire, & consequently of heating it. So that if the vis viva communicated to the electric fluid in the wire is sufficiently great the wire will receive a violent heat. But though the vis viva communicated to the <electric fluid in the wire> should not be sufficient to produce any sensible increase of total heat, yet it seems not impossible that the particles of the wire may be displaced thereby in such manner as to cause a great diminution of latent heat & consequently a great increase of sensible heat

It will be proper therefore to examine what degree of vis viva can be given to the electric fluid by the discharge, in doing which I shall argue upon the principles laid down in my paper concerning the cause of electricity*

Let ACB & ADE be 2 wires communicating at A with the positive side of an electric jar, while the negative side communicates with the ground; & let the 2 wires touch each other at B. They will become overcharged & will repel each other with a

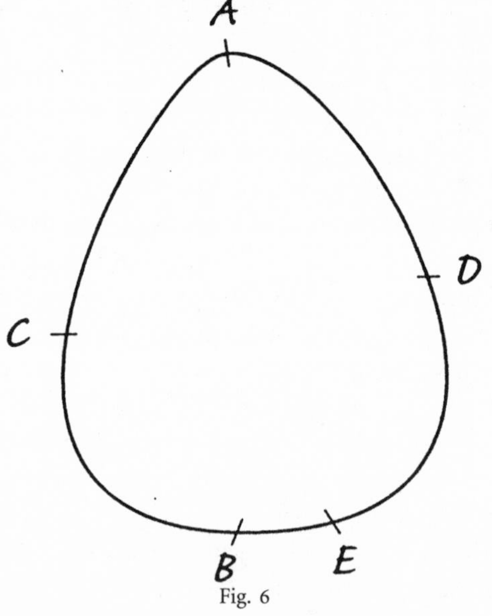

Fig. 6

viva cannot take place without displacing some of their particles, an alteration of latent heat will take place, which will commonly make the alteration of sensible heat very different from what it would otherwise be

*Phil. trans Vol 61 P. 584

force equal to the repulsion of the redundant fluid in ACB on the redundant fluid in ADB. Take now the point E such that the repulsion of the wire ACB on that part of the wire ADB which is between A & E shall be small in comparison of its repulsion on the part BE; so that the repulsion of the wires may be considered as equal to the repulsion of the redundant fluid in ACB on the redundant fluid in BE. Let this distance be called d. Let the repulsion of the 2 wires be equal to the weight of a piece of wire BE whose length is r; & let the redundant fluid on the positive side of the jar, the redundant fluid in BE & the whole quantity of fluid in BE be equal in weight to pieces of the same wire whose lengths are g, f & ϕ respectively. Let now the wire ADB be removed; & let its end A be made to communicate with the negative side of the jar; & then let the end B be suddenly brought in contact with the same end B of the other wire, so as to discharge the jar. Now the redundant fluid on the positive side of the jar is equal to the whole quantity of fluid containd in a piece of the wire whose length is $d \times \frac{g}{\phi}$ so that the fluid in the wire will move in consequence of the discharge through the space $\frac{dg}{\phi}$. Moreover each particle of fluid in BE is repelled from B with the same force that the particles of redundant fluid were repelled in the former position of the wire, & are therefore repelled with a force which is to their weight as r to f; & the whole force with which they are repelled is equal to the weight of a piece of the wire whose length equals $\phi \times \frac{r}{f}$. Therefore the velocity which this fluid acquires by the discharge is that which a body acquires by gravity in falling from the height $\frac{gd}{\phi} \times \frac{r}{f}$; & the vis viva given to it is the same which a piece of the wire whose length $= \phi$ acquires by falling from the height $\frac{gdr}{f\phi}$, or which a piece whose length is $\frac{gd}{f}$ acquires by falling from the height r

It must be observed that the vis viva given to the electric fluid in the whole length of wire is no more than what is given to that in BE; for as the fluid in the part ADE is impelled only by that in BE, it can receive no vis viva without taking away as much from that in BE. Moreover $\frac{gd}{f}$ is the length of a piece of wire of the same thickness as BE, which if electrified with the same force as the jar, would receive as much redundant fluid as is collected on the positive side of the jar; therefore if this length is called l, the whole vis viva communicated to the electric fluid in the wire, is that which a piece of the wire whose length is l, acquires in falling from the height r

It should seem therefore that the vis viva given to the electric fluid by the discharge is by no means sufficient to account for the heat produced, except by displacing the particles & thereby causing a diminution of latent heat This must be examined[14]

The force with which the electric fluid in BE is impelled was before said to be equal to the weight of a piece of wire whose length is $\frac{r\phi}{f}$; & therefore <is most likely> is not unlikely to be very great; as the quantity of redundant fluid in an electrified body is <most likely> not unlikely to be very small in comparison of the whole quantity

The velocity given to this fluid was before said to be that acquired by falling from the height $\frac{dg}{f} \times \frac{r}{\phi}$, or $l \times \frac{r}{\phi}$; & therefore if the weight of the electric fluid naturally containd in a body is excessively small in proportion to the whole weight of the body, as is very likely to be the case, the velocity given to the electric fluid will be very great, but not otherwise; so that it is uncertain whether this velocity is great or not

It has been commonly supposed that the velocity of the electric fluid is excessively great; but there is nothing which shews whether it is or not. It has been found indeed that if a jar is discharged through a very long wire interrupted in the middle & each end, there is no sensible interval of time between the appearance of the spark at the middle & at the ends. But this shews nothing as to the velocity with which the fluid moves; for we are not to suppose that the same fluid which issues from the positive side of the jar, passes through the whole length of the wire, & enters in at the negative side. On the contrary the space through which the electric fluid which issues from the jar moves is in all probability not great; but it serves to push forwards the electric fluid before it & thereby to propagate the motion through the wire, just as the motion of the particles of air propagate sound; & the swiftness with which the motion is propagated through the wire does not at all depend on the velocity of the electric fluid, any more than the velocity of sound depends on that with which the particles of air vibrate

On the expansion of bodies by heat & of the different forms which they assume according to their heat

As the particles of which bodies consist are in perpetual vibration, it is evident that the attraction & repulsion of any 2 particles on each other must be perpetually varying according to the part of the vibration they are in; & even their mean attraction & repulsion can hardly be the same as if they were at rest in their mean position; so that an increase of heat in a body can hardly fail of being attended with an alteration in the distance & arrangement of its particles; & therefore it may be expected that an increase of heat in a body should cause some alteration in its bulk. Why this alteration is always in excess, & why the size of the body is never diminished thereby, I do not pretend to explain; but there seems no reason why it may not be so

It may also be expected that when the vibrations of the particles are sufficiently great, their mutual attractions & repulsions may be so much alterd, as to oblige the particles to assume a quite different arrangement, so as totally to alter the appearance & properties of the body; as is the case in evaporation & the melting & hardening of bodies

It may be observed that in general bodies grow more fluid or less hard, & their particles have a less strong adhesion to each other, as their heat increases. It may also be observed that most chymical decompositions & combinations are promoted by heat; both of which seem no unnatural effects of an increased vibration of the particles

Conclusion

It has been shewn therefore by as strict reasoning as can be expected in subjects not purely mathematical, that if heat consists in the vibrations of the particles of bodies, the effects will be strikingly analogous, & as far as our experiments yet go, in no case contradictory to the phenomena. For first bodies must retain their heat without increase or diminution until alterd, either by receiving heat from or communicating it to other bodies, or by some other external cause. 2ndly If 2 bodies of different heats are placed in contact, one will communicate heat to the other till they both acquire the same sensible heat, that is till they shew the same heat by the thermometer. 3rd It is reasonable to expect that different quantities of heat should be required to communicate the same sensible heat to different bodies, & that the chymical union of different bodies or a change in the nature of any body, should be commonly attended with a change of sensible heat. 4th It is proved that no heat should be produced by light reflected from or transmitted through a body, but only by the light absorbed by it. 5th It is rendered probable that the emission of light & rays of heat should not take place without some degree of heat, & should be increased as the heat increases. 6th Bodies should receive some degree of heat from friction & hammering; but in all probability it can hardly be sensible except when a violent force is impressed on a small quantity of matter & in all probability the sensible heat given to different bodies by the same degree of friction should be different, & that more[over] this should proceed from their different capacities for heat. All these circumstances agree perfectly with the phenomena. <the phenomena of the heating of bodies by friction & hammering nowise disagree but rather agree with the theory> 7th It is shewn to be probable that an increase of heat in a body ought to be attended, by an alteration either of its bulk or some other of its properties or both; & that a sufficient increase of heat should intirely alter the nature of the body The heat also produced by electricity seems by no means inconsistent with the theory though it is an effect which I should not have expected; nor do I know of any other phenomenon which is inconsistent with it; though the theory is not of that pliable nature as to be easily adapted to any appearances. On the other hand the various hypotheses which have been formed for explaining the phenomena of heat by a fluid seem to shew that none of them are very satisfactory; & though it does not seem impossible that a fluid might exist endued with such properties as to produce the effects of heat; yet any hypothesis of such kind must be of that unprecise nature, as not to admit of being reduced to strict reasoning, so as to suffer one to examine whether it will really explain the phenomena or whether it will not rather be attended with numberless inconsistencies & absurdities. So that though it might be natural for philosophers to adopt such an hypothesis when no better offerd itself; yet when a theory has been proposed by S[r] I[saac] N[ewton] which, as may be shewn by strict reasoning, must produce effects strongly analogous to those observed to take place, & which seems no ways inconsistent with any, there can no longer be any reason for adhering to the former hypothesis

But to put the matter in a stronger light. It seems certain that the action of such rays of light as are absorbed by a body must produce a motion & vibration of its particles; so that it seems certain that the particles of bodies must actually be in motion; & this motion must produce effects analogous to most of the phenomena of heat, & seems to disagree with none. Why therefore should we have recourse to the hypothesis of a fluid, which nothing proves the existence of, when a circumstance which certainly does exist, seems fully sufficient to account for the phenomena

Again; it seems certain, as was before said, that the particles of bodies must have an internal motion, & that this motion must produce effects which can hardly fail of manifesting themselves to our senses; but no phenomena occur except those of heat, which can with any probability be attributed to this cause; which is another strong argument for supposing that heat <must> consists in this motion of the particles of bodies

[Changes in the Paper]

Additions & alterations

P. 1) Pr. 1) The increase of vis viva which a body already in motion acquires in moving through a given space by a given force is equal to the whole vis viva which it would acquire in moving from rest through the same space [This is inserted in the paper, as directed, p. 176.]

ENDNOTES FOR PREFACE, INTRODUCTION, AND PART ONE

Preface

1. Henry Cavendish, *The Electrical Researches of the Honourable Henry Cavendish*, ed. J. C. Maxwell (Cambridge: Cambridge University Press, 1879).

2. Henry Cavendish, *The Scientific Papers of the Honourable Henry Cavendish, F.R.S.*, vol. 1, *The Electrical Researches*, ed. J. C. Maxwell, rev. J. Larmor; vol. 2, *Chemical and Dynamical*, ed. E. Thorpe, C. Chree, F. W. Dyson, A. Geikie, and J. Larmor (Cambridge: Cambridge University Press, 1921).

3. In the possession of John Charles Compton Cavendish, 5th Baron Chesham.

4. Part 4, "Henry Cavendish's Scientific Letters," in Christa Jungnickel and Russell McCormmach, *Cavendish, the Experimental Life* (Lewisburg, PA: Bucknell University Press, 1999), 515–731.

5. Letter from Dietrich Belitz, 23 Apr. 2001.

Introduction

1. "Science," *Encyclopaedia Britannica* (Edinburgh, 1797) 16:705. Charles Hutton, *A Mathematical and Philosophical Dictionary*. . . . new ed., 2 vols. (London, 1815) 2: 183. Samuel Vince, "On the Motion of Bodies Affected by Friction," *PT* 75 (1785): 165–89, on 165.

2. Maurice Crosland and Crosbie Smith, "The Transmission of Physics from France to Britain: 1800–1840," *Historical Studies in the Physical Sciences* 9 (1978): 1–61, on 9.

3. I. B. Cohen, *Franklin and Newton: An Inquiry into Speculative Newtonian Experimental Science and Franklin's Work in Electricity as an Example Thereof* (Philadelphia: American Philosophical Society, 1956), 351.

4. Richard Sorrenson, "Towards a History of the Royal Society in the Eighteenth Century," *Notes and Records of the Royal Society of London* 50 (1996): 29–46, on 37.

5. John Locke, *An Essay Concerning Human Understanding*, ed. A. C. Fraser, vol. 2 (Oxford: Clarendon Press, 1894), 244, 249.

6. David Hume, *A Treatise of Human Nature*, ed. L. A. Selby-Bigge (Oxford: Oxford University Press, 1955), 448.

7. P. M. Heimann and J. E. McGuire, "Newtonian Forces and Lockean Powers: Concepts of Matter in Eighteenth-Century Thought," *Historical Studies in the Physical Sciences* 3 (1971): 233–306, on 284.

8. James Hutton, *An Investigation of the Principles of Knowledge, and of the Progress of Reason, from Sense to Science and Philosophy*, 3 vols. (Edinburgh, 1794) 1:518–19; 2:256–57, 262, 266, 269, 277, 329.

9. George Adams, *Lectures on Natural and Experimental Philosophy*, 5 vols. (London, 1794) 1:62.

10. [Charles Blagden], Obituary of Henry Cavendish, *Gentleman's Magazine*, March 1810, 292.

11. Frank Manuel, *A Portrait of Isaac Newton* (Cambridge, MA: Harvard University Press, 1968), 388–90.

12. Robert K. Merton, *On the Shoulders of Giants: A Shandean Postscript* (San Diego: Harcourt Brace Jovanovich, 1985). N. J. Thrower, ed., *Standing on the Shoulders of Giants: A Longer View of Newton and Halley. Essays Commemorating the Tercentenary of Newton's* Principia *and the 1985–1986 Return of Comet Halley* (Los Angeles: University of California Press, 1990).

13. Galileo Galilei, *Dialogue Concerning the Two Chief World Systems*, trans. S. Drake (Los Angeles: University of California Press, 1967), 23, 33, 130; *Dialogues Concerning Two New Sciences*, trans. H. Crew and A. de Salvio (New York: McGraw-Hill, 1963), 1.

14. Johannes Kepler, *Epitome of Copernican Astronomy and Harmonies of the World*, trans. C. G. Wallis (Amherst: Prometheus Books, 1995), 17, 55, 57, 90.

15. Galileo, *Dialogue*, 113.

16. Albert Einstein, "Foreword," in Sir Isaac Newton, *Opticks; or A Treatise of the Reflections, Refractions, Inflections and Colours of Light*, 4th ed. of 1730 (New York: Dover, 1952), lix–lx, on lix.

17. J. T. Desaguliers, *System of Experimental Philosophy* (London, 1719), 2.

18. Charles Morton, "Observations and Experiments upon Animal Bodies, Digested in a Philosophical Analysis, or Inquiry into the Cause of Voluntary Muscular Motion," *PT* 47 (1751): 305–14, on 305.

19. Barbara Lovett Cline, *The Questioners*, published by Thomas Y. Crowell in 1965; reprinted, *Men Who Made a New Physics: Physicists and the Quantum Theory* (New York: New American Library, 1969).

20. Albert Einstein, "Autobiographical Notes," *Albert Einstein: Philosopher-Scientist*, 2 vols., ed. P. A. Schilpp (La Salle: Open Court, 1949) 1:1–94, on 17.

21. Quoted in Richard Rhodes, *The Making of the Atomic Bomb* (New York, London, Toronto: Simon & Schuster, 1986), 77.

22. Simon Schaffer, "Scientific Discoveries and the End of Natural Philosophy," *Social Studies of Science* 16 (1986): 387–420.

23. Robert Symmer, "New Experiments and Observations Concerning Electricity," *PT* 51 (1759): 340–93, on 348.

24. Albert Einstein, "Foreword," in Galileo, *Dialogue*, vi–xix, on xvii–xviii.

25. Albert Einstein, "Isaac Newton," *Out of My Later Years* (Totowa, NJ: Littlefield, Adams, 1967), 201–4, on 201.

26. Albert Einstein, "Principles of Theoretical Physics," *Ideas and Opinions*, trans. S. Bargmann (New York: Dell, 1954), 216–19, on 219.

27. Thomas Young, *A Course of Lectures on Natural Philosophy and the Mechanical Arts*, 2 vols. (London, 1807) 1:7.

28. John Playfair, *Illustrations of the Huttonian Theory of the Earth* (Edinburgh, 1802), 524–25.

29. Dugald Stewart, *The Collected Works of Dugald Stewart, Esq., F.R.SS.*, ed. W. Hamilton, 11 vols. (Edinburgh, 1854–1860), 3:329.

30. Murray Gell-Mann, *The Quark and the Jaguar: Adventures in the Simple and the Complex* (New York: W. H. Freeman, 1994), 77.

31. Ronald N. Giere, *Explaining Science: A Cognitive Approach* (Chicago: University of Chicago Press, 1998), 62.

32. Some of these meanings are given in Ernest Nagel, "Theory and Observation," in Ernest Nagel, Sylvain Bromberger, and Adolf Grünbaum, *Observation and Theory in Science* (Baltimore: The Johns Hopkins University Press, 1971), 21–22.

33. Hume, *Treatise of Human Nature*, 450.

34. These reasons concern the satisfactions that the pursuit of astronomy and natural philosophy brings, according to Roger Long, *Astronomy, in Five Books*, 2 vols. in 3 (Cambridge, 1742–84) 1:vii–viii, 2:585–86, 725.

35. Charles Blagden to "My Lord" [Lord George Cavendish], draft. n.d. [April 1810], Blagden Collection, Royal Society, Misc. Matter—Unclassified.

36. Russell McCormmach, "Henry Cavendish on the Theory of Heat," *Isis* 79 (1988): 37–67. Jungnickel, Christa, and Russell McCormmach, *Cavendish, the Experimental Life* (Lewisburg, PA: Bucknell University Press, 1999).

37. A. Rupert Hall, *From Galileo to Newton 1630–1720* (London: Collins, 1963), 321.

38. Cohen, *Franklin and Newton*, 223–24.

39. P. M. Heimann, "Newtonian Natural Philosophy and the Scientific Revolution," *History of Science* 11 (1973): 1–7, on 1, 4.

Chapter 1

1. Charles Hutton, *A Mathematical and Philosophical Dictionary . . .*, 2 vols. (London, 1795–96) 2:227. Hutton took this statement from a book by the philosopher George Berkeley in 1710, *A Treatise Concerning the Principles of Human Knowledge*, ed. C. P. Krauth (Philadelphia: J. B. Lippincott, 1881), 251–52.

2. Stewart, *Collected Works* 3:242–44.

3. William Enfield, *Institutes of Natural Philosophy, Theoretical and Experimental.* . . . (London, 1785), vii.

4. Thomas L. Hankins, *Science and the Enlightenment* (Cambridge: Cambridge University Press, 1985), 7.

5. Colin Maclaurin, *An Account of Sir Isaac Newton's Philosophical Discoveries, in Four Books.* . . . (London, 1748), 10.

6. John Locke, *An Essay Concerning Human Understanding*, abr., ed. A. S. Pringle-Pattison (Oxford: Clarendon Press, 1950), 3. Thomas Reid, *The Works of Thomas Reid, D.D.*, ed. W. Hamilton, 7th ed., vol. 2 (Edinburgh, 1872), 768.

7. Wesley C. Salmon, *Scientific Explanation and the Causal Structure of the World* (Princeton, NJ: Princeton University Press, 1984), 240, 259.

8. This section is taken partly from Jungnickel and McCormmach, *Cavendish, the Experimental Life*, 169–71.

9. Jungnickel and McCormmach, *Cavendish, the Experimental Life*, 113.

10. I acknowledge a helpful discussion of these points by Bruce Hevly at Johns Hopkins University.

11. Papers were not classified by subject in the original *Philosophical Transactions*, but they were in an abridged edition. The largest class was natural philosophy, including astronomy, called "mechanical philosophy"; the next largest was chemistry, meteorology, and geology, called "chemical philosophy"; the others were natural history, physiology, and mathematics. These headings had long been in use in the abridgments.

12. Henry Cavendish, "Experiments to Determine the Density of the Earth," *PT* 88 (1798): 469–526.

13. Charles Bazerman, *Shaping Written Knowledge: The Genre and Activity of the Experimental Article in Science* (Madison: University of Wisconsin Press, 1988), 66–68.

14. William Watson, "A Collection of the Electrical Experiments Communicated to the Royal Society," *PT* 45 (1748): 49–120.

15. John L. Greenberg, "Mathematical Physics in Eighteenth-Century France," *Isis* 77 (1986): 59–78, on 77.

16. Crosland and Smith, "Transmission of Physics from France to Britain, 5–9.

17. Ibid., 55, 60.

18. Ibid., 55, 59–60.

19. David Philip Miller, "The Revival of the Physical Sciences in Britain, 1815–40," *Osiris*, 2d ser. 2 (1986): 107–34, on 134.

20. Bruce Hevly, "Afterword: Reflections on Big Science and Big History," *Big Science: The Growth of Large-Scale Research*, ed. P. Galison and B. Hevly (Stanford, CA: Stanford University Press, 1992), 355–63, on 357, 361.

Chapter 2

1. Gerd Buchdahl, *The Image of Newton and Locke in the Age of Reason* (London: Sheed and Ward, 1961), 2, 21.

2. John Locke, *An Essay Concerning Human Understanding*, abr. ed. A. S. Pringle-Pattison (Oxford: Clarendon Press, 1950), 66–67.

3. David Hume, *A Treatise of Human Nature*, 226–31.

4. Hutton, *Investigation of the Principles of Knowledge* 1:345, 358–59.

5. Matthew Young, *An Analysis of the Principles of Natural Philosophy* (Dublin, 1800), 1.

6. Hutton, *Investigation of the Principles of Knowledge* 1:345, 358–59.

7. "Physics," "Physiology," *Encyclopaedia Britannica* (Edinburgh, 1777) 3:478. Hutton, *Dictionary* 2:229. Note the title of Benjamin Martin, *The Philosophical Grammar; Being a View of the Present State of Experimental Physiology, or Natural Philosophy*, 5th ed. (London, 1755).

8. In the eighteenth century, the word "science" was sometimes restricted to deductive reasoning. That was the meaning given to it in Thomas Parkinson, *A System of Mechanics, Being the Substance of Lectures upon That Branch of Natural Philosophy* (Cambridge, 1785), 8. William Ludlam stated that by reasoning from abstract, general ideas, which are acquired not from the senses but from the operation of the mind, knowledge is extended to a great many particular subjects; "this makes what is called science." *The Rudiments of Mathematics; Designed for the Use of Students at the Universities: Containing an Introduction to Algebra, Remarks on the First Six Books of Euclid, the Elements of Plain Trigonometry* (Cambridge, 1785), 137, 140. Natural philosophers for much of our period could not have meant what we do by "science," Andrew Cunningham claims, since science as an "intensional" practice was not invented until c. 1780–c. 1850. "Getting the Game Right: Some Plain Words on the Identity and Invention of Science," *Studies in the History and Philosophy of Science* 19 (1988): 365–89, on 385.

9. There was no agreement on what constituted the "natural." Hume identified three meanings of the word, and he used at least two more implicitly, all characterized by oppositions: nature as opposed to the miraculous, the unusual, the artificial, the civic, and the moral. Hume, *Treatise of Human Nature*, 474–75, including editor's note.

10. Hutton, *Dictionary* 2:139.

11. Examples: Robert Young, *An Essay on the Powers and Mechanism of Nature; Intended by a Deeper Analysis of Physical Principles, to Extend, Improve, and More Firmly Establish, the Grand Superstructure of the Newtonian System* (London, 1788), quotation from the preface; William Jones, *An Essay on the First Principles of Natural Philosophy: Wherein the Use of Natural Means, or Second Causes, in the Oeconomy*

of the Material World, Is Demonstrated from Reason, Experiments of Various Kinds, and the Testimony of Antiquity . . . In Four Books. . . . (Oxford, 1762); Simon Schaffer, "Natural Philosophy," *The Ferment of Knowledge*, ed. G. S. Rousseau and R. S. Porter (Cambridge: Cambridge University Press, 1980), 55–91, on 55–71.

12. Simon Schaffer, "Natural Philosophy and Public Spectacle in the Eighteenth Century," *History of Science* 21 (1983): 1–43; "Scientific Discoveries and the End of Natural Philosophy"; "Natural Philosophy," 73, 81, 83, 86–87.

13. Hutton, *Dictionary*, new ed., 2:183. Parkinson, *Mechanics*, 2.

14. Schaffer, "Natural Philosophy," 85.

15. J. A. Deluc, *An Elementary Treatise on Geology: Determining Fundamental Points in That Science, and Containing an Examination of Some Modern Geological Systems, and Particularly of the Huttonian Theory of the Earth*, trans. H. de la Fite (London, 1809), iv.

16. [John Robison], "Physics," *Encyclopaedia Britannica*, 3d ed. (Edinburgh, 1797) 14:637–59, on 647. Crosbie Smith, " 'Mechanical Philosophy' and the Emergence of Physics in Britain: 1800–1850," *Annals of Science* 33 (1976): 3–29, on 8.

17. William Lewis, *Commercium Philosophico-Technicum; or, The Philosophical Commerce of Arts: Designed as an Attempt to Improve Arts, Trades, and Manufactures* (London, 1763), iii–v.

18. For example, Thomas Young, *A Syllabus of a Course of Lectures on Natural and Experimental Philosophy* (London, 1802).

19. [Robison], "Physics," 647.

20. Tiberius Cavallo, *The Elements of Natural or Experimental Philosophy*, 4 vols. (London, 1803) 2:7–13.

21. John Rowning, *A Compendious System of Natural Philosophy: With Notes Containing Mathematical Demonstrations, and Some Occasional Remarks*, 3d ed. (London, 1738), preface.

22. Anthony Shepherd, *The Heads of a Course of Lectures in Experimental Philosophy Read at Christ College* (Cambridge, [1770]), 3–4.

23. Samuel Vince, *The Heads of a Course of Lectures on Experimental Philosophy; Comprising All the Fundamental Principles in Mechanics, Hydrostatics, and Optics; with an Explanation of the Construction and Use of All the Principal Instruments in Astronomy; Together with Magnetism and Electricity* (Cambridge, 1795). George Atwood, *An Analysis of a Course of Lectures on the Principles of Natural Philosophy, Read in the University of Cambridge. . . .* (London, 1784).

24. Hutton, *Dictionary* 2:229.

25. Adams, *Natural and Experimental Philosophy*, 5 vols. (London, 1794) 1:126, 129.

26. William Nicholson, *A Dictionary of Chemistry . . .*, 2 vols. in 1 (London, 1795), v.

27. Isaac Newton, *Sir Isaac Newton's Mathematical Principles of Natural Philosophy*

and His System of the World, trans. A. Motte in 1729, ed. F. Cajori, 2 vols. (Los Angeles: University of California Press, 1962) 2:398–400.

28. Cavallo, *Natural or Experimental Philosophy* 1:5.

29. Richard Feynman, *The Character of Physical Laws* (New York: Modern Library, 1994), 162, 167.

30. Rowning, *Natural Philosophy*, 167.

31. Thomas Reid, quoted in L. L. Laudan, "Thomas Reid and the Newtonian Turn of British Methodological Thought," *The Methodological Heritage of Newton*, ed. R. E. Butts and J. W. Davis (Toronto: University of Toronto Press, 1970), 103–31, on 110.

32. Hutton, *Investigation of the Principles of Knowledge* 1:521, 529.

33. Colin Maclaurin, *Account*, 12.

34. Adam Smith, "The Principles Which Lead and Direct Philosophical Inquiries; Illustrated by the History of Astronomy," *The Whole Works of Adam Smith, LL.D. F.R.S. &c.*, new ed., vol. 5 (London, 1822), 1–80, on 16.

35. Roger Long, *Astronomy* 1:198.

36. Oliver Goldsmith, *A Survey of Experimental Philosophy, Considered in Its Present State of Improvement*, 2 vols. (London, 1766) 1:19. Adams, *Natural and Experimental Philosophy* 3:1.

37. C. B. Wilde, "Hutchinsonianism, Natural Philosophy and Religious Controversy in Eighteenth Century Britain," *History of Science* 18 (1980): 1–24, on 17. Larry Stewart, "Seeing Through the Scholium: Religion and Reading Newton in the Eighteenth Century," *History of Science* 34 (1996): 123–64, on 124, 138, 140, 142, 156.

38. Isaac Newton, *Opticks; or A Treatise of the Reflections, Refractions, Inflections and Colours of Light*, 4th ed. of 1730 (New York: Dover, 1952), 405; *Principia* 2:544–46.

39. Maclaurin, *Account*, 4, 6, 17, 22–23.

40. Adams, *Natural and Experimental Philosophy* 1, preface.

41. Heimann and McGuire, "Newtonian Forces and Lockean Powers," 281, 292, 305. P. M. Heimann. " 'Nature is a Perpetual Worker': Newton's Aether and Eighteenth-Century Natural Philosophy," *Ambix* 20 (1973): 1–25, on 24. A Cambridge lecturer on natural philosophy in the middle of the century denied that gravity needs "God's continual Energy," for that implies an imperfect Creation, leading to a "very mean opinion of God." Newton had confronted the same objection. Lectures read by Gervas Holmes in Emmanuel College around 1757, in the handwriting of Richard Watson, later Professor of Chemistry and Bishop of Llandoff. "Mr. Holmes's Lectures," Cambridge University Library, p. 53.

42. Richard Watson, "Observations on the Sulphur Wells at Harrowgate, Made in July and August, 1785," *PT* 76 (1786): 171–88, on 188.

43. Thomas Rutherforth, *A System of Natural Philosophy; Being a Course of Lectures in Mechanics, Optics, Hydrostatics, and Astronomy, Which Are Read in St. Johns College Cambridge . . .*, 2 vols. (Cambridge, 1748) 1:7–9.

44. Andrew Cunningham maintains that natural philosophy was about God, and that when nature was studied for itself it was no longer natural philosophy. Given Cavendish's well-known disinterest in religion, if he had a spiritual leaning, we will probably never know. "How the *Principia* Got Its Name: Or, Taking Natural Philosophy Seriously," *History of Science* 29 (1991): 377–92, on 381, 388; "Getting the Game Right," 384.

45. Colin Maclaurin, *A Treatise of Fluxions, in Two Books*, 2 vols. (Edinburgh, 1742) 1:53.

46. Newton, *Principia* 1:6–10. Benjamin Martin, *Philosophia Britannica; or, A New and Comprehensive System of the Newtonian Philosophy, Astronomy and Geography. In a Course of Twelve Lectures. . . .*, 2 vols. (Reading, 1747) 1:47–48. Bechler, "Newton's Ontology of the Force of Inertia," 287–304, on 301–2. Thomas Young questioned the meaningfulness of absolute motion, and denied that we can determine absolute rest. *Natural Philosophy and the Mechanical Arts*, (London, 1807) 1:18–20.

47. Adams, *Natural and Experimental Philosophy* 3:2–3.

48. Ibid.

49. William Nicholson, *An Introduction to Natural Philosophy*, 2 vols. (London, 1782) 1:7–17. Cavallo, *Natural or Experimental Philosophy* 1:21. Parkinson, *Mechanics*, 34.

50. William L. Harper, "Reasoning from Phenomena: Newton's Argument for Universal Gravitation and the Practice of Science," *Action and Reaction: Proceedings of a Symposium to Commemorate the Tercentenary of Newton's* Principia, ed. P. Theeman and A. F. Seeff (Newark, NJ: University of Delaware Press, 1993), 144–82, on 167–68.

51. P. M. Heimann and J. E. McGuire, "Cavendish and the *Vis Viva* Controversy: A Leibnizian Postscript," *Isis* 62 (1970): 225–27; "Newtonian Forces and Lockean Powers."

52. Joseph Priestley, *Disquisitions Relating to Matter and Spirit. To Which Is Added, the History of the Philosophical Doctrine Concerning the Origin of the Soul, and the Nature of Matter . . .* (London, 1777), 7–19. Joseph Priestley to Joseph Bretland, 7 Mar. 1773, *A Scientific Autobiography of Joseph Priestley (1733–1804): Selected Scientific Correspondence*, ed. R. E. Schofield (Cambridge, MA: MIT Press, 1966), 116–18, on 117.

53. Joseph Priestley, *The History and Present State of Discoveries Relating to Vision, Light, and Colours*, 2 vols. (London, 1772) 2:391–94. Hutton, *Dictionary* 2:83. Parkinson, *Mechanics*, 40. Thomas Young, *Natural and Experimental Philosophy*, 145. Heimann and McGuire, "Newtonian Forces and Lockean Powers," 236–37, 240, 276–79.

54. Martin, *Philosophia Britannica*, preface.

55. Cavallo, *Natural or Experimental Philosophy* 1:11–13; 2:15, 515–19.

56. Jones, *Natural Philosophy*, 5, 13.

57. Cavallo, *Natural or Experimental Philosophy* 1:1.

58. Thomas Hornsby, "Dr. Hornsby's 1st Lecture of the Properties of Bodies," MSS. Rigaud, Bodleian Library, University of Oxford, 26254.54, fol. 188.

59. Locke, *Essay*, abr. ed., 153.

60. George Berkeley, *The Principles of Human Knowledge*, 1710, and *De Motu*, 1721, discussed in Ernan McMullin, "Enlarging the Known World," *Physics and Our View of the World*, ed. J. Hilgevoord (Cambridge, 1994) 95.

61. Hume, *Treatise of Human Nature*, 86–89.

62. David Hume, *An Enquiry Concerning Human Understanding*, ed. T. L. Beauchamp (Oxford: Oxford University Press, 1999), 109.

63. John Passmore, "Hume, David," *Dictionary of Scientific Biography*, ed. C. C. Gillispie, vol. 6 (New York: Charles Scribner's Sons, 1972), 555–60, on 557–58.

64. S. A. Grave, *The Scottish Philosophy of Common Sense* (Oxford: Clarendon Press, 1960), 4.

65. Heimann and McGuire, "Newtonian Forces and Lockean Powers," 267.

66. Grave, *Scottish Philosophy*, 136–37. Laudan, "Thomas Reid," 106–7, 124–25, 128–29. Edward H. Madden, "The Reidian Tradition: Growth of the Causal Concept," *Nature and Scientific Method*, ed. D. O. Dahlstrom (Washington, DC: The Catholic University of America Press, 1991), 291–307, on 293–95.

67. Stewart, *Collected Works* 2:6.

68. John Playfair, *Outlines of Natural Philosophy, Being Heads of Lectures Delivered in the University of Edinburgh*, 2 vols. in 1 (Edinburgh, 1812–14) 1:3.

69. John Robison, *A System of Mechanical Philosophy . . .*, ed. D. Brewster, 4 vols. (Edinburgh, 1822) 1:3.

70. Newton made the impressed force proportional to the change of motion, or momentum, not as we do to its rate of change.

71. R. W. Home, "The Third Law in Newton's Mechanics," *British Journal for the History of Science* 4 (1968): 39–51, on 41, 46.

72. Newton, *Principia* 1:13–14.

73. Ibid., 1:192.

74. George Fordyce, "The Croonian Lecture on Muscular Motion," *PT* 78 (1788): 23–36, on 36.

75. Heimann and McGuire, "Newtonian Forces and Lockean Powers," 302.

76. Stephen Hales, *Vegetable Staticks . . .* (London, 1727).

77. A. L. Donovan, *Philosophical Chemistry in the Scottish Enlightenment: The Doctrines and Discoveries of William Cullen and Joseph Black* (Edinburgh: Edinburgh University Press, 1975), 21–28. Robert E. Schofield, *Mechanism and Materialism: British Natural Philosophy in an Age of Reason* (Princeton, NJ: Princeton University Press, 1970), 74–75.

78. J. T. Desaguliers, "Some Thoughts and Conjectures Concerning the Cause of Elasticity," *PT* 41 (1739): 175–85, on 175.

79. Gowin Knight, *An Attempt to Demonstrate, That All the Phaenomena in Nature May Be Explained by Two Simple Active Principles, Attraction and Repulsion: Wherein the Attractions of Cohesion, Gravity, and Magnetism, Are Shewn to Be One and the Same; and the Phaenomena of the Latter Are More Particularly Explained* (London, 1748).

80. Newton, *Principia* 1:2.

81. Bechler, "Newton's Ontology of the Force of Inertia," 293–94.

82. Peter Harman, "Concepts of Inertia: Newton to Kant," *Religion, Science, and Worldview: Essays in Honor of Richard S. Westfall*, ed. M. J. Osler and P. L. Farber (Cambridge: Cambridge University Press, 1985), 109–33, on 124. Dudley Shapere, "The Philosophical Significance of Newton's Science," *The* Annus Mirabilis *of Sir Isaac Newton*, ed. R. Palter (Cambridge, MA: MIT Press, 1970), 285–99, on 288–90.

83. Matthew Young, *Natural Philosophy*, 26–27.

84. Newton, *Principia* 1:2–3, 40–42. Adams, *Natural and Experimental Philosophy* 1:2–3. Robison, *Mechanical Philosophy* 1:6–14. I. Bernard Cohen, "Newton's Second Law and the Concept of Force in the *Principia*," *The* Annus Mirabilis *of Sir Isaac Newton*, ed. R. Palter (Cambridge, MIT Press, 1970), 143–85, on 144. James E. McGuire, "Comment," ibid., 186–91, on 187. Thomas L. Hankins, "The Reception of Newton's Second Law of Motion in the Eighteenth Century,"*Archives Internationales d'Histoire des Sciences* 20 (1967): 43–65, on 46–51.

85. Fordyce, "The Croonian Lecture on Muscular Motion," 29.

86. William Emerson, *The Principles of Mechanics. Explaining and Demonstrating the General Laws of Motion, the Laws of Gravity, Motion of Descending Bodies . . .*, 2d ed. (London, 1758), iii.

87. Bryan Robinson, *A Dissertation on the Aether of Sir Isaac Newton* (Dublin, 1743); *Sir Isaac Newton's Account of the Aether, with Some Additions by Way of Appendix* (Dublin, 1745).

88. Thomas Young, *Natural and Experimental Philosophy*, 145.

89. Newton, *Opticks*, 395.

90. Roger Joseph Boscovich, *A Theory of Natural Philosophy*, trans. J. M. Child from the 2d ed. of 1763 (Cambridge, MA: MIT Press, 1966), 22.

91. Adams, *Natural and Experimental Philosophy* 4:472.

92. Playfair, *Natural Philosophy* 2:340–41.

93. Steven Weinberg, *Dreams of a Final Theory* (New York: Pantheon Books, 1992), 8–12, 17–18. Albert Einstein, "The Fundaments of Theoretical Physics," *Out of My Later Years* (Totowa, NJ: Littlefield, Adams, 1967), 95–107, on 96.

94. Desaguliers, *A System of Experimental Philosophy*, 1.

95. As in the title of Cavallo, *The Elements of Natural or Experimental Philosophy*.

96. Maclaurin, *Account*, 90–91.

97. W. White, "Experiments upon Air; and the Effects of Different Kinds of Effluvia upon It; Made at York," *PT* 68 (1778): 194–220, on 203.

98. Hutton, *Dictionary* 1:458.

99. Sorrenson, "Towards a History of the Royal Society," 37.

100. Tiberius Cavallo, *A Complete Treatise of Electricity in Theory and Practice; with Original Experiments (London, 1777); A Treatise on Magnetism in Theory and Practice; with Original Experiments* (London, 1787).

101. Cohen, *Franklin and Newton,*" 189. Henry Guerlac, "Newton and the Method of Analysis," *Essays and Papers in the History of Modern Science* (Baltimore: The Johns Hopkins University Press, 1977), 193–215, on 212. Thomas L. Hankins, "Newton's 'Mathematical Way' a Century after the *Principia*," *Some Truer Method: Reflections on the Heritage of Newton*, ed. F. Durham and R. D. Purrington (New York: Columbia University Press, 1990), 89–112, on 96, 108.

102. Playfair, *Illustrations*, 527.

103. Newton, *Principia* 1:xvii–xviii.

104. Goldsmith, *Experimental Philosophy* 1:18.

105. John Harris, *Lexicon Technicum*, 5th ed. (London, 1736). Hutton, *Dictionary* 2:157. Cohen, *Franklin and Newton*, 180–81.

106. Richard S. Westfall, *Science and Religion in Seventeenth-Century England* (Ann Arbor: University of Michigan Press, 1973), 72.

107. Ernan McMullin, *Newton on Matter and Activity* (Notre Dame, IN: University of Notre Dame Press, 1978), 73–74.

108. Jones, *Natural Philosophy*, 5. George Horne, *A Fair, Candid, and Impartial Account of the Case between Sir Isaac Newton and Mr. Hutchinson . . .* (Oxford, 1753), 57–59. John C. English, "John Hutchinson's Critique of Newtonian Heterodoxy," *Church History* 68 (1999): 581–97.

109. Martin, *Philosophia Britannica*, 13.

110. Charles Henry Wilkinson, *An Analysis of a Course of Lectures on the Principles of Natural Philosophy, to Which Is Prefixed, An Essay on Electricity, with a View of Explaining the Phenomena of the Leyden Phial, etc. on Mechanical Principles* (London, 1799). G. N. Cantor, "Anti-Newton," *Let Newton Be!* ed. J. Fauvel, R. Flood, M. Shorthand, and R. Wilson (Oxford: Oxford University Press, 1988), 203–22, on 215.

111. Robison, *Mechanical Philosophy*. Matthew Young, *Natural Philosophy*, 10. Smith, " 'Mechanical Philosophy,' " 4, 8.

112. Playfair, *Natural Philosophy*, 3–4.

113. Thomas Young, *Natural Philosophy* 1:13.

114. Newton, *Opticks, 176. John Roche, "Newton's Principia,"* *Let Newton Be!* ed. J. Fauvel, R. Flood, M. Shortland, and R. Wilson (Oxford: Oxford University Press, 1988), 43–62, on 57.

115. William Charles Wells, *Essay upon Single Vision with Two Eyes* (London, 1792), 26.

116. Joseph Priestley, *The History and Present State of Electricity, with Original Experiments* (London, 1767), 443.

117. Parkinson, *Mechanics*, 6.

118. Nicholson, *Natural Philosophy* 1:6.

119. Richard Kirwan, "Thoughts on Magnetism," *TRIA* 6 (1797): 177–91, on 177.

120. Mark Akenside, "Observations on the Origin and Use of the Lymphatic Vessels of Animals . . . ," *PT* 50 (1757): 322–28, on 325–26.

121. William Irvine, *Essays, Chiefly on Chemical Subjects*, ed. W. Irvine, Jr. (London, 1805), 171.

122. Feynman, *Character of Physical Laws*, 34. Gell-Mann, *The Quark and the Jaguar*, 108.

123. Maclaurin, *Account*, 8.

124. Guerlac, "Newton and the Method of Analysis," 211.

125. McMullin, *Newton on Matter and Activity*, 43, 47.

126. Newton, *Principia* 1:xvii; 2:397.

127. Hutton, *Dictionary* 2:229; new ed., 2:23–24.

128. I. B. Cohen, "Commentary by I. Bernard Cohen," *Scientific Change: Historical Studies in the Intellectual, Social and Technical Conditions for Scientific Discovery and Technical Invention, from Antiquity to the Present*, ed. A. C. Crombie (New York: Basic Books, 1961): 466–71, on 470–71.

129. Newton, *Principia* 1:4–6. William Ludlam, "Remarks on Sir Isaac Newton's Second Law of Motion," 4 May 1780, Journal Book of the Royal Society, Royal Society, 30:34–38, on 35.

130. John Hunter, "Of the Heat, etc, of Animals and Vegetables," *PT* 68 (1778): 7–49. Charles Hutton, "The Force of Fired Gun-Powder, and the Initial Velocities of Cannon Balls, Determined by Experiments; from Which Is Also Deduced the Relation of the Initial Velocity of the Weight of the Shot and the Quantity of Powder," *PT* 68 (1778): 50–85.

131. J. Ravetz, "The Representation of Physical Quantities in Eighteenth-Century Mathematical Physics," *Isis* 52 (1961): 7–20, on 9.

132. Playfair, *Natural Philosophy* 1:2. Robison, *Mechanical Philosophy* 1:19–20. Thomas Reid, "An Essay on Quantity; Occasioned by Reading a Treatise in Which Simple and Compound Ratios Are Applied to Virtue and Merit, by the Rev. Mr. Reid; Communicated in a Letter from the Rev. Henry Miles D.D. and F. R. S. to Martin Folkes Esq; Pr.R. S.," *PT* 45 (1748): 505–20, on 506. This essay was written by Reid in response to an application of algebra to morals by Francis Hutcheson. The author of the paper in the *Philosophical Transactions* is somewhat unclear, because in the journal, "Communicated in a Letter from" can mean "by." Formerly I mistook the author to be Miles, not Reid. The citation is corrected in the Second Printing, 2001,

of Jungnickel and McCormmach, *Cavendish, the Experimental Life*, 177–78. Thomas Reid, "An Essay on Quantity; Occasioned by Reading a Treatise in Which Simple and Compound Ratios Are Applied to Virtue and Merit," *The Works of Thomas Reid, D.D.*, 2: 715–19.

133. Reid, "Essay on Quantity," 511.

134. Robert Smith, *A Compleat System of Optics in Four Books, viz. a Popular, a Mathematical, a Mechanical, and a Philosophical Treatise. To Which Are Added Remarks upon the Whole*, 2 vols. (Cambridge, 1738) 1:6.

135. Robert Smith, *Harmonics, or The Philosophy of Musical Sounds* (Cambridge, 1749; reprint, New York: Da Capo, 1966), 248.

136. Colin Maclaurin, *A Treatise of Algebra, in Three Parts . . .* (London, 1748), 298.

137. Maclaurin, *Fluxions* 1:52; *Account*, 92–93.

138. John Stewart, commentary, in Newton, *Sir Isaac Newton's Two Treatises*, vii–x.

139. William Emerson, *The Doctrine of Fluxions: Not Only Explaining the Elements Thereof, but Also Its Application and Use in the Several Parts of Mathematics and Natural Philosophy*, 3d ed. (London, 1768), xiii.

140. Cavallo, *Natural or Experimental Philosophy* 1:10.

141. Emerson, *Fluxions*, iii–iv. William Ludlam, *Mathematical Essays*, 2d ed. (Cambridge, 1787), 63–64. Maclaurin, *Fluxions* 1:52–53, 56–57. Differing from other authors on the subject, Thomas Simpson rejected the view of fluxions as "meer velocities," in *The Doctrine and Application of Fluxions*, 2 vols. (London, 1750) 1:vi. Benjamin Robins, *A Discourse Concerning the Nature and Certainty of Sir Isaac Newton's Methods of Fluxions and of Prime and Ultimate Ratios;* in *Mathematical Tracts of the Late Benjamin Robins, Esq.*, ed. J. Wilson, Vol. 1 (London, 1761), 7–77, on 32–35, 44.

142. [Robison], "Physics," 654.

143. Benjamin Wilson, address to Pierre Moultou, Jr., in *A Short View of Electricity* (London, 1780).

144. J. F. Scott, "Maclaurin, Colin," *Dictionary of Scientific Biography*, ed. C. C. Gillispie, vol. 8 (New York: Charles Scribner's Sons, 1973), 609–12, on 610–11. I. Grattan-Guinness, "French *Calcul* and English Fluxions around 1800: Some Comparisons and Contrasts," *Jahrbuch Überblicke Mathematik* (Mannheim: Bibliographisches Institut, 1986), 167–78, on 167–68.

145. Emerson, *Fluxions*, 354, 380.

146. Goldsmith, *Experimental Philosophy* 1:14–15.

147. Sorrenson, "Towards a History of the Royal Society," 37.

148. John Robison spoke of the law of electric force as a mathematical "function" of distance, which he determined by experiment to be the inverse square of the distance. *Mechanical Philosophy* 4:67, 74.

149. Emerson, *Fluxions*, xv.

150. Benjamin Martin, *A New and Comprehensive System of Mathematical Insti-*

tutions, Agreeable to the Present State of the Newtonian Mathesis, 2 vols. (London, 1759–64) 1:386.

151. Thomas S. Kuhn, "Mathematical versus Experimental Traditions in the Development of Physical Science," *The Essential Tension: Selected Studies in Scientific Tradition and Change* (Chicago: University of Chicago Press, 1977), 31–65, on 61.

152. Henry Cavendish, "An Attempt to Explain Some of the Principal Phaenomena of Electricity, by Means of an Elastic Fluid," *PT* 61 (1771): 584–677; in *Electrical Researches,* 3–63, on 10, 15, 17, 19, 28.

153. A. Wolf, *A History of Science, Technology and Philosophy in the 18th Century,* ed. D. McKie, 2 vols. (New York: Harper & Brothers, 1961) 1:56–58.

154. Martin, *Mathematical Institutions* 1:361.

155. Thomas Simpson, *Miscellaneous Tracts . . .* (London, 1757), preface.

156. For instance, in the translation of a French mathematical paper on the precession of the equinoxes, *PT* 48 (1754): 385–441.

157. Maclaurin, *Fluxions* 2:420.

158. W. W. Rouse Ball, *A History of the Study of Mathematics at Cambridge* (Cambridge: Cambridge University Press, 1889), 72–73, 98.

159. Roger Penrose et al., *The Large, the Small and the Human Mind* (Cambridge: Cambridge University Press, 1997), 2–3, 169.

160. Gerard 't Hooft, "Questioning the Answers or Stumbling upon Good and Bad Theories of Everything," *Physics and Our View of the World,* ed. J. Hilgevoord (Cambridge: Cambridge University Press, 1994), 16–37, on 34.

Chapter 3

1. Albert Einstein, "The Mechanics of Newton and Their Influence on the Development of Theoretical Physics," *Ideas and Opinions,* 247–55, on 247–48.

2. Newton, *Opticks,* 334.

3. Roger Cotes, *Hydrostatical and Pneumatical Lectures,* ed. R. Smith (London, 1738), 123.

4. Playfair, *Illustrations,* 187.

5. Maclaurin, *Account,* 10.

6. Hales quoted in Stewart, *Collected Works* 3:303.

7. Rev. Mr. [G. C.?] Morgan, "Observations and Experiments on the Light of Bodies in a State of Combustion," *PT* 75 (1785): 190–212, on 190–91, 207–8.

8. William Watson, "A Sequel to the Experiments and Observations Tending to Illustrate the Nature and Properties of Electricity," *PT* 44 (1747): 704–49, on 743–49. Priestley, *History and Present State of Electricity,* 449.

9. Alexander Wilson, "Observations on the Solar Spots," *PT* 64 (1774): 1–30, on 17–18, 21, 28–29.

10. Alexander Wilson, "An Answer to the Objections Stated by M. De la Lande, in the Memoirs of the French Academy in the Year 1776, against the Solar Spots Being Excavations in the Luminous Matter of the Sun, together with a Short Examination of the Views Entertained by Him upon That Subject," *PT* 73 (1783): 144–68, on 160–65.

11. Alexander Wilson, *Thoughts on General Gravitation, and Views Thence Arising as to the State of the Universe* (London, 1777), 5–7.

12. John Goodricke, "A Series of Observations on, and a Discovery of, the Period of the Variation of the Light of the Bright Star in the Head of Medusa, Called Algol," *PT* 73 (1783): 474–82, on 482.

13. Edward Pigott, "Observations of a New Variable Star," *PT* 75 (1785): 127–36, on 134–35.

14. Locke, *Essay*, 2 ed. 379–80.

15. Hugh Hamilton, *Four Introductory Lectures in Natural Philosophy* (London, 1774), 6–8. Playfair, *Illustrations*, 514.

16. Maclaurin, *Account*, 4–8.

17. Newton, *Principia* 2:547.

18. Newton, *Opticks* 1: 404–5.

19. Barry Gower, *Scientific Method: An Historical and Philosophical Introduction* (New York: Routledge, 1997), 79.

20. Robison, *Mechanical Philosophy* 4:41.

21. William Morgan, "Electrical Experiments Made in Order to Ascertain the Non-conducting Power of a Perfect Vacuum, etc.," *PT* 75 (1785): 272–78, on 278.

22. Edward King, "An Attempt to Account for the Universal Deluge," *PT* 57 (1767): 44–57, on 44.

23. Sir William Herschel, "On the Proper Motion of the Sun and Solar System; with an Account of Several Changes That Have Happened among the Fixed Stars since the Time of Mr. Flamstead," *PT* 73 (1783): 247–83, on 248, 268, 275.

24. Martin, *Philosophia Britannica* 1: preface. Alexander Wilson, "An Answer to the Objections," 164.

25. Maclaurin, *Account*, 346.

26. Henry Eeles, "Concerning the Cause of the Ascent of Vapour and Exhalation, and Those of Winds; and of the General Phaenomena of the Weather and Barometer," *PT* 49 (1755): 124–54, on 124–25.

27. I. B. Cohen, *Franklin and Newton* 138–41, 174.

28. Joseph Priestley, "An Account of Further Discoveries in Air," *PT* 65 (1775): 384–94, on 389.

29. Hugh Hamilton, *Philosophical Essays on the Following Subjects: I. On the Principles of Mechanics. II. On the Ascent of Vapours . . . III. Observations and Conjectures on the Nature of the Aurora Borealis, and the Tails of Comets* (Dublin, 1766), 92.

30. Quoted from Isaac Milner's lectures in Cambridge in 1782, shortly before he was appointed Jacksonian Professor. L. J. M. Coleby, "Isaac Milner and the Jacksonian Chair of Natural Philosophy," *Annals of Science* 10 (1954): 234–57, on 242.

31. Robison quoted and discussed by Robinson M. Yost, "Pondering the Imponderable: John Robison and Magnetic Theory in Britain (c. 1775–1805)," *Annals of Science* 56 (1999): 143–74, on 166.

32. Eeles, "Concerning the Cause of the Ascent of Vapour," 124–25.

33. Stephen Hales quoted in Stewart, *Collected Works* 3:303.

34. John Pringle, "Some Remarks upon the Several Accounts of the Fiery Meteor (Which Appeared on Sunday 26th of November, 1758), and upon Other Such Bodies," *PT* 51 (1760): 259–74, on 273.

35. Three-page printed directions, dated 6 Nov. 1783, *A Plan for Observing the Meteors Called Fire-Balls.* Nevil Maskelyne to William Herschel, 3 Nov. 1783, Royal Astronomical Society, Mss Herschel, W 1/13, M26. Charles Blagden to Sir Joseph Banks, 16 Oct. 1783, Fitzwilliam Museum Library, Percival H189 and H190. Charles Blagden, "An Account of Some Late Fiery Meteors: with Observations," *PT* 74 (1784): 201–32, on 222–32.

36. Hutton, *Dictionary* 1:621.

37. Ibid.

38. Paul Feyerabend, *Against Method* (London: NLB, 1975).

39. Erasmus Darwin, *Zoonomia; or The Laws of Organic Life*, 2 vols. (London, 1794–96) 1:2.

40. William Nicholson, translator's preface to the second edition, with notes from the French, of Richard Kirwan, *An Essay on Phlogiston, and the Constitution of Acids*, 2d ed. (London, 1789), iii.

41. Newton, *Principia* 1:xix.

42. Einstein, "Principles of Theoretical Physics," 217.

43. Gell-Mann, *The Quark and the Jaguar*, 92–93.

44. Maclaurin, *Account*, 3.

45. Priestly, *Vision, Light, and Colours*, 270.

46. Hutton, *Dictionary*, new ed., 2:505.

47. William Henly, "An Account of Some New Experiments in Electricity . . . ," *PT* 64 (1774): 389–431, on 400, 406; "Experiments and Observations in Electricity," *PT* 67 (1777): 85–143, on 99–101, 104, 107.

48. Thomas Wright, *An Original Theory or New Hypothesis of the Universe, Founded upon the Laws of Nature, and Solving by Mathematical Principles the General Phaenomena of the Visible Creation; and Particularly the Via Lactae . . .* (London, 1750).

49. Stewart, *Collected Works* 3:302, 307.

50. Cavallo, *Treatise of Electricity*, 109; *Treatise on Magnetism*, 106.

51. Priestley, *History and Present State of Electricity*, 444–45.

52. Henry Cavendish, "On the Height of the Luminous Arch Which Was Seen on Feb. 23, 1784," *PT* 80 (1790): 101–5, on 104.

53. Playfair, *Natural Philosophy*, 3.

54. John Robison, "On the Motion of Light, as Affected by Refracting and Reflecting Substances, Which Are Also in Motion," *TRSE* 2:2 (1788): 83–111, on 95–96.

55. John Freke, *An Essay to Shew the Cause of Electricity; and Why Some Things Are Non-electricable . . . In a Letter to Mr. William Watson, F. R. S.* (London, 1746), viii.

56. Nicholson, *Dictionary*, v.

57. James Keir, "Experiments and Observations on the Dissolution of Metals and Acids, and Their Precipitations; with an Account of a New Compound Acid Menstruum, Useful in Some Technical Operations of Parting Metals," *PT* 80 (1790): 359–84, on 361.

58. Playfair, *Illustrations*, 527.

59. Samuel Horsley, "M. De Luc's Rules, for the Measurement of Heights by the Barometer, Compared with Theory . . . ," *PT* 64 (1774): 214–303, on 218.

60. Playfair, *Illustrations*, 511–12, 526.

61. Priestley, *History and Present State of Electricity*, 445–46.

62. George Atwood, *A Treatise on the Rectilinear Motion and Rotation of Bodies; with a Description of Original Experiments Relative to the Subject* (Cambridge, 1784), iv–v, 123, 151–55, 294.

63. Erasmus Darwin, "Remarks on the Opinion of Henry Eeles, Esq; Concerning the Ascent of Vapour, Published in the Philosoph. Transact. Vol. XLIX. Part I, P. 124," *PT* 50 (1757): 240–54, on 248–49.

64. James Hutton, *A Dissertation upon the Philosophy of Light, Heat, and Fire. In Seven Parts* (Edinburgh, 1794), 4.

65. Reid, 1780, quoted in Laudan, "Thomas Reid," 124.

66. William Ludlam, *Astronomical Observations Made in St. John's College, Cambridge, in the Years 1767 and 1768: With an Account of Several Astronomical Instruments* (Cambridge, 1769), preface.

67. George Atwood, "Investigations, Founded on the Theory of Motion, for Determining the Times of Vibration of Watch Balances," *PT* 84 (1794): 119–68, on 120.

68. Tiberius Cavallo, "Of the Methods of Manifesting the Presence, and Ascertaining the Quality, of Small Quantities of Natural or Artificial Electricity," *PT* 78 (1788): 1–22, on 22.

69. George Atwood, "A General Theory for the Mensuration of the Angles Subtended by Two Objects, of Which One Is Observed by Rays after Two Reflections from Plane Surfaces, and the Other by Rays Coming Directly to the Spectator's Eye," *PT* 71 (1781): 395–435, on 396.

70. William Herschel, "On the Parallax of the Fixed Stars," *PT* 72 (1782): 82–111, on 104–5.

71. Jan Ingen-Housz, "Some Farther Considerations on the Influence of the Vegetable Kingdom on the Animal Creation," *PT* 72 (1782): 426–39, on 429.

72. Hutton, *Dictionary* 2:183.

73. Henry Ussher, "Account of the Observatory Belonging to Trinity College, Dublin," 1 *TRIA* (1787): 3–22, on 5.

74. Preface to *TRIA* 1 (1787): ix–xvii, on xii.

75. Alan Q. Morton, "Concepts of Power: Natural Philosophy and the Uses of Machines in Mid-Eighteenth-Century London," *British Journal for the History of Science* 28 (1995): 63–78, on 64–65.

76. David Philip Miller, "The Usefulness of Natural Philosophy: The Royal Society and the Culture of Practical Utility in the Later Eighteenth Century," *British Journal for the History of Science* 32 (1999): 185–201, on 200.

77. *The Oxford Universal Dictionary on Historical Principles*, ed. C. T. Onions, 3d ed. (Oxford: Clarendon Press, 1955), 694.

78. Playfair, *Illustrations*, 511–12, 526.

79. Darwin, "Remarks on the Opinion of Henry Eeles," 248.

80. A. R. Humphreys, "The Literary Scene," *From Dryden to Johnson*, vol. 4 of the Pelican Guide to English Literature, ed. B. Ford (Baltimore: Penguin Books, 1963), 51–93, on 75.

81. William Blackstone, *Commentaries on the Laws of England*, 13th ed. (London, 1800). "Fictions," *Encyclopaedia Britannica*, vol. 9 (Chicago: William Benton, 1962), 218.

82. Hume, *Treatise of Human Nature*, 197, 209.

83. Jeremy Bentham, *Bentham's Theory of Fictions*, ed. C. K. Ogden (New York: Harcourt, Brace, 1932), lxiii.

84. Ibid., 18.

85. Roy Porter, *Enlightenment: Britain and the Creation of the Modern World* (London: Allen Lane, 2000), 289.

86. [Benjamin Robins], "An Account of a Book Intitled, New Principles of Gunnery, Containing the Determination of the Force of Gunpowder; and an Investigation of the Resisting Power of the Air to Swift and Slow Motions . . . as Far as the Same Relates to the Force of Gunpowder," *PT* 42 (1743): 437–56, on 445–46.

87. [Robins], "Account," 446–47, 454.

88. Cavendish, "An Attempt to Explain Some of the Principal Phaenomena of Electricity, 584–677; in *Electrical Researches*, 3-63, on 29, 39. Maxwell's proof that Cavendish's supposition was justified is given in *Electrical Researches*, 375–76.

89. Adams, *Natural and Experimental Philosophy*, 3:10.

90. Barnes, Bloor, and Henry, *Scientific Knowledge*, 107–8. Giere, *Explaining Science*, 79–85.

91. I. B. Cohen, "The *Principia*, the Newtonian Style, and the Newtonian Revolution in Science," *Action and Reaction: Proceedings of a Symposium to Commemorate the Tercentenary of Newton's* Principia, ed. P. Theerman and A. F. Seeff (Newark, NJ: University of Delaware Press, 1993), 61–104, on 95.

92. C. Truesdell, *The Rational Mechanics of Flexible or Elastic Bodies 1638–1788*, Editor's Introduction to *Leonhardi Euleri Opera omnia*, 2d ser., vol. 11 (Zürich: Orell Füssli, 1960), 139.

93. Robison, *Mechanical Philosophy* 1:317

94. John Smeaton, "An Experimental Inquiry Concerning the Natural Powers of Water and Wind to Turn Mills, and Other Machines, Depending on Circular Motion," *PT* 51 (1760): 100–74, on 100–1.

95. Benjamin Thompson, "On the Propagation of Heat in Fluids," *The Complete Works of Count Rumford*, 4 vols. (Boston, 1870–75) 1:239–400, on 376.

96. Quoted in K. Wilber, ed., *Quantum Questions: Mystical Writings by the World's Great Physicists* (Boston: Shambhala, 1985), 9.

97. Cavallo, *Natural or Experimental Philosophy* 1:17.

98. [Robison], "Physics," 648.

99. J. T. Desaguliers, "Some Things Concerning Electricity," *PT* 41 (1740): 634–37, on 634.

100. Stewart, *Collected Works* 2:7.

101. Symmer, "New Experiments and Observations," 388.

102. Martin, *Mathematical Institutions* 2:8. Hutton, *Dictionary* 2:139–40. Richard Helsham, *A Course of Lectures in Natural Philosophy* (London, 1739), 34.

103. Emerson, *Mechanics*, 4.

104. Henry Cavendish, "Plan of a Treatise on Mechanicks," Cavendish Mss VIb, 45:17.

105. Reid, *Works*, 607.

106. James Jurin, "An Inquiry into the Measure of the Force of Bodies in Motion: With a Proposal of an Experimentum Crucis, to Decide the Controversy about It," *PT* 43 (1745): 423–40, on 426.

107. John Ellicott, "Several Essays towards Discovering the Laws of Electricity," *PT* 45 (1748): 195–224, on 203. Joseph Priestley, "Experiments Relating to Phlogiston, and the Seeming Conversion of Water into Air," *PT* 73 (1783): 398–434, on 399, 429. Herschel, "On the Proper Motion of the Sun and Solar System," 276.

108. Cohen, "The *Principia*," 70–71.

109. Jones, *Essay*, 13.

110. James Hall, "Account of a Series of Experiments, Shewing the Effects of Compression in Modifying the Action of Heat," *TRSE* 6 (1805): 71–184, on 75.

111. Henry Cavendish, "Experiments on Air," *PT* 74 (1784): 119–53, on 150, 152.

112. John Lyon, *Experiments and Observations Made with a View to Point Out the*

Errors of the Present Received Theory of Electricity; and Which Tend in Their Progress to Establish a New System, on Principles More Conformable to the Simple Operations of Nature (London, 1780), 12.

113. Vince, "On the Motion of Bodies Affected by Friction," 165, 177.

114. Cavendish, "An Attempt to Explain Some of the Principal Phaenomena of Electricity," 659, 666, 670.

115. Ibid., 42–43.

116. Ibid., 8–9.

117. Henry Cavendish, "[Experiments on the Charges of Bodies]," *Electrical Researches*, 114–43, on 142.

118. Playfair, *Natural Philosophy* 1:4.

119. Newton, *Principia* 1:xvii–xviii; *Opticks*, 404.

120. J. T. Desaguliers, *Experimental Philosophy*, 2 vols. (London, 1734–44) 1:346–56.

121. William Herschel, "Catalogue of the Second Thousand of New Nebulae and Clusters of Stars; with a Few Introductory Remarks on the Construction of the Heavens," *PT* 79 (1789): 212–55, on 212–14.

122. William Herschel, "On the Construction of the Heavens," *PT* 75 (1785): 213–66, on 213–14, 217, 220.

123. Herschel, "Catalogue of the Second Thousand of New Nebulae and Clusters of Stars," 219–21.

124. Cavallo, "Small Quantities of Natural or Artificial Electricity," 2.

125. John Michell, "Conjectures Concerning the Cause, and Observations upon the Phaenomena, of Earthquakes . . . ," *PT* 51 (1760): 566–634, on 569.

126. Newton, *Opticks*, 100, 102.

127. Patrick Murdoch, "Rules and Examples for Limiting the Cases in Which the Rays of Refracted Light May Be Reunited into a Colourless Pencil," *PT* 53 (1763): 173–94, on 186.

128. Matthew Young, "Demonstration of Newton's Theorem for the Correction of Spherical Errors in the Object Glasses of Telescopes," *TRIA* 4 (1790): 171–75.

129. Newton, *Opticks*, cxxi. Robert Blair, "Experiments and Observations on the Unequal Refrangibility of Light," *TRSE* 3:2 (1791): 3–76, on 68–69.

130. Priestley, *Vision, Light, and Coulors* 1:357.

131. Newton, *Opticks*, 370.

132. Henry John Steffens, *The Development of Newtonian Optics in England* (New York: Science History Publications, 1977), 48–53. G. N. Cantor, *Optics after Newton: Theories of Light in Britain and Ireland, 1704–1840* (Manchester: Manchester University Press, 1983), 204–5.

133. John Michell, "On the Means of Discovering the Distance, Magnitude, &c. of the Fixed Stars, in Consequence of the Diminution of the Velocity of Their Light, in Case Such a Diminution Should Be Found to Take Place in Any of Them, and

Such Other Data Should Be Procured from Observations, as Would Be Farther Necessary for That Purpose," *PT* 74 (1784): 35–57, on 37, 51.

134. James Hall, "Experiments on Whinstone and Lava," *TRSE* 5:1 (1798): 43–75, on 44–45; "Account of a Series of Experiments, Shewing the Effects of Compression in Modifying the Action of Heat," 75–76.

135. Thomas Melvill, "Discourse Concerning the Cause of the Different Refrangibilities of the Rays of Light," *PT* 48 (1753): 261–70, on 262.

136. Samuel Vince to William Herschel, 10 Jan. 1784, Royal Astronomical Society, Herschel Mss W/13 V.2.

137. Patrick Wilson, "An Experiment Proposed for Determining, by the Aberration of the Fixed Stars, Whether the Rays of Light, in Pervading Different Media, Change Their Velocity According to the Law Which Results from Sir Isaac Newton's Ideas Concerning the Cause of Refraction; and for Ascertaining Their Velocity in Every Medium Whose Refractive Density Is Known," *PT* 72 (1782): 58–70, on 59.

138. Robison, "On the Motion of Light," 84, 90, 95–98.

139. Samuel Horsley, "Difficulties in the Newtonian Theory of Light, Considered and Removed," *PT* 60 (1770): 417–40.

140. Abraham Bennet, "A New Suspension of the Magnetic Needle, Intended for the Discovery of Minute Quantities of Magnetic Attraction: Also an Air Vane of Great Sensibility; with New Experiments on the Magnetism of Iron Filings and Brass," *PT* 82 (1792): 81–98, on 87–88.

141. Thomas Young, "Outlines of Experiments and Inquiries Respecting Sound and Light," *PT* 90 (1800): 106–50; "The Bakerian Lecture. On the Theory of Light and Colours," *PT* 92 (1802): 12–48, on 44, 46, 48.

142. G. N. Cantor, "Was Thomas Young a Wave Theorist?" *American Journal of Physics* 52 (1984): 305–8.

143. Papacino D'Antoni, *A Treatise on Gunpowder; A Treatise on Fire-Arms; and A Treatise on the Service of Artillery in the Time of War*, trans. Captain Thomas of the Royal Regiment of Artillery (London, 1789), xvii; quoted in Brett D. Steele, "Muskets and Pendulums: Benjamin Robins, Leonhard Euler, and the Ballistics Revolution," *Technology and Culture* 35 (1994): 348–82, on 348.

144. Quoted in Steele, "Muskets and Pendulums," 348.

145. Address by the President, Martin Folkes, on presenting the Copley Medal to Robins, 1 Dec. 1746, Journal Book of the Royal Society 19:160–65, on 164.

146. Charles Hutton, *Tracts on Mathematical and Philosophical Subjects . . .*, 3 vols. (London, 1812) 3:209; *Dictionary* 1:563; 2:292.

147. Hutton, *Dictionary* 2:287.

148. Robins, *Mathematical Tracts*, 59–76. Robins built upon Daniel Bernoulli's analysis of shot propelled by compressed air and by gunpowder. Bernoulli recognized that the work done on the shot by the expanding air or by the ignited gunpowder

is equal to one half the vis viva of the shot. Daniel Bernoulli and Johann Bernoulli, *Hydrodynamics by Daniel Bernoulli and Hydraulics by Johann Bernoulli*, trans. T. Carmody and H. Kobus (New York: Dover, 1968), 265. Steele, "Muskets and Pendulums," 358.

149. Hutton, *Tracts on Mathematical and Philosophical Subjects* 3:210; *Dictionary* 1:564–65. Steele, "Muskets and Pendulums," 360.

150. According to Martin Folkes, 1 Dec. 1746, Journal Book of the Royal Society 19:161–62.

151. Hutton, "Force of Fired Gun-Powder," 51–52; *Dictionary* 1:565, 570; *Tracts* 3: 216.

152. Benjamin Thompson, "New Experiments upon Gun-Powder, with Occasional Observations and Practical Inferences; to Which Are Added, an Account of a New Method of Determining the Velocities of All Kinds of Military Projectiles, and the Description of the Very Acurate Eprouvette for Gun-Powder," *PT* 71 (1781): 229–328, on 298, 305, 307. Hutton, *Dictionary* 1:568.

153. Benjamin Thompson, "Experiments to Determine the Force of Fired Gunpowder," *PT* 87 (1797): 222–92, on 224, 232–34.

154. Richard Kirwan, *An Estimate of the Temperature of Different Latitudes* (London, 1787), iii.

155. William Borlase, quoted in J. Oliver, "William Borlase's Contributions to Eighteenth-Century Meteorology and Climatology," *Annals of Science* 25 (1969): 275–317, on 291.

156. Hutton, *Dictionary* 2:677.

157. Playfair, *Illustrations*, 2, 513–16.

158. Priestley, *History and Present State of Electricity*, 160, 449–50, 455, 458, 480.

159. Symmer, "New Experiments and Observations Concerning Electricity," 371, 387–89.

160. Thomas Harding, "Observations on the Variation of the Needle," *TRIA* 4 (1790): 107–18, on 107.

161. J. Lorimer, "Description of a New Dipping-Needle," *PT* 65 (1775): 79–84, 80.

162. Bennet, "A New Suspension of the Magnetic Needle," 96–98.

163. Long, *Astronomy* 2: 717.

Chapter 4

1. Some sections of this chapter are taken from part 3, chapter 5, "Mercury," in Jungnickel and McCormmach, *Cavendish, the Experimental Life*, 393–423.

2. Adams, *Lectures on Natural and Experimental Philosophy* 1:245.

3. J. L. Heilbron, *Electricity in the 17th and 18th Centuries: A Study of Early Modern Physics* (Los Angeles: University of California Press, 1979), 81.

4. Newton's paper on the subject was published in the *Philosophical Transactions* in 1701, and appended to Cotes, *Hydrostatical and Pneumatical Lectures*, 213–23.

5. Newton, *Principia* 2:522. George Martine, *Essays and Observations on the Construction and Graduation of Thermometers, and on the Heating and Cooling of Bodies*, 2d ed. (Edinburgh, 1772).

6. Newton, *Opticks*, 339.

7. Ibid., 339–45, 395.

8. Ibid., 371, 375–406.

9. Ibid., 349.

10. Ibid., 379–80, 399.

11. Madden, "The Reidian Tradition," 304. Smith, " 'Mechanical Philosophy,' " 7–8. Richard Olson, *Scottish Philosophy and British Physics 1750–1880: A Study in the Foundations of the Victorian Scientific Style* (Princeton, NJ: Princeton University Press, 1975), 44–47, 57. Donovan, *Philosophical Chemistry*, 58–59. [Robison], "Physics."

12. Cullen's paper was first published in *Essays and Observations, Philosophical and Literary, Read before a Society in Edinburgh and Published by Them* (Edinburgh, 1756), and republished together with Black's essay, *Experiments upon Magnesia Alba, Quick-Lime, and Other Alkaline Substances; by Joseph Black. To Which Is Annexed, An Essay on the Cold Produced by Evaporating Fluids, and of Some Other Means of Producing Cold; by William Cullen* (Edinburgh, 1777), 115–33, quotation on 132.

13. Douglas McKie and Niels H. de V. Heathcote, *The Discovery of Specific and Latent Heats* (London: Arnold, 1935), 15–20, quote on p. 6. Henry Guerlac, "Black, Joseph," *Dictionary of Scientific Biography*, ed. C. C. Gillispie, vol. 2 (New York: Charles Scribner's Sons, 1970), 173–83, on 177. Donovan, *Philosophical Chemistry*, 238–240.

14. Donovan, *Philosophical Chemistry*, 240–46.

15. McKie and Heathcote, *Discovery*, 13, 35.

16. Ibid., 35, 122–23.

17. Donovan, *Philosophical Chemistry*, 265–71.

18. Irvine, *Essays*, 115–27, 188.

19. Robert Fox, *The Caloric Theory of Gases from Lavoisier to Regnault* (Oxford: Clarendon, 1971), 31–32.

20. Black's lectures were edited by his pupil John Robison, in Edinburgh in 1803. An American publication soon followed: Joseph Black, *Lectures on the Elements of Chemistry*, 3 vols., ed. J. Robison (Philadelphia, 1806–7).

21. E. L. Scott, "Richard Kirwan, J. H. de Magellan, and the Early History of Specific Heat," *Annals of Science* 38 (1981): 141–53, on 150–51. T. H. Lodwig and W. A. Smeaton, "The Ice Calorimeter of Lavoisier and Laplace and Some of Its Critics," *Annals of Science* 31 (1974): 1–18, on 4.

22. Adair Crawford, *Experiments and Observations on Animal Heat, and the Inflammation of Combustible Bodies. Being an Attempt to Resolve These Phaenomena into a General Law of Nature* (London, 1779), 12–15. William Cleghorn, "William Cleghorn's *De Igne* (1779)," ed. and trans. D. McKie and N. H. de V. Heathcote, *Annals of Science* 14 (1958): 1–82. P. Dugud Leslie, *A Philosophical Inquiry into the Cause of Animal Heat: With Incidental Observations on Several Physiological and Chymical Questions, Connected with the Subject* (London, 1778). J. H. Magellan, *Essai sur la Nouvelle Théorie du Feu Elémentaire, et de la Chaleur des Corps* (London, 1780). Donovan, *Philosophical Chemistry*, 272–73.

23. This is according to the experimental lectures of the Jacksonian Professor in Cambridge, Isaac Milner. In successive years, 1784–87, he delivered to the Vice Chancellor of the University an essay on heat based on his lectures. Milner's "Essay," MS. Ee 5.35, Cambridge University Library, pp. 14, 19. Also quoted from in Coleby, "Isaac Milner and the Jacksonian Chair of Natural Philosophy," 242–43.

24. Leslie, *Philosophical Inquiry into the Cause of Animal Heat*, 221ff.

25. Black, *Lectures* 1:31–32. Donovan, *Philosophical Chemistry*, 229–30.

26. Irvine, *Essays*, 187. Scott, "Richard Kirwan, J. H. de Magellan," 145–46.

27. Cleghorn, "*De Igne*."

28. George Fordyce, "An Account of Some Experiments on the Loss of Weight in Bodies on Being Melted or Heated," *PT* 75 (1785): 361–65, on 364; "An Account of an Experiment on Heat," *PT* 77 (1787): 310–17, on 316.

29. Benjamin Thompson, "An Inquiry Concerning the Weight Ascribed to Heat," *PT* 89 (1799): 179–94, on 180, 192.

30. Discussion of heat as a fluid: Black, *Lectures* 1:31ff.

31. Fox, *Caloric Theory*, 19–25.

32. Schofield, *Mechanism and Materialism*, 157–90. P. M. Heimann, "Ether and Imponderables," *Conceptions of Ether: Studies in the History of Ether Theories, 1740–1900*, ed. G. N. Cantor and M. J. S. Hodge (Cambridge: Cambridge University Press, 1981), 61–83, on 67–73. J. L. Heilbron, *Weighing Imponderables and Other Quantitative Science around 1800*, Supplement to *Historical Studies in the Physical and Biological Sciences*, vol. 24, pt. 1 (Berkeley: University of California Press, 1993), 5–33.

33. Black, *Lectures* 1:80.

34. Fox, *Caloric Theory*, 19–25.

35. Milner, "Essay," 24–25, 33. Coleby, "Milner," 245–52.

36. Newton, *Principia* 1:xvii.

37. Parkinson, *Mechanics*, 31–32.

38. Cavallo, *Natural or Experimental Philosophy*, 1:40.

39. Robison, *Mechanical Philosophy* 1:261.

40. For example: George Atwood, *An Analysis of a Course of Lectures on the Principles of Natural Philosophy, Read in the University of Cambridge . . .* (London,

1784); Samuel Vince, *A Plan of a Course of Lectures on the Principles of Natural Philosophy* (Cambridge, 1793); Robison, *Mechanical Philosophy.*

41. Emerson, *Mechanics,* iv.

42. Cavallo, *Natural or Experimental Philosophy* 1:42.

43. Smith, " 'Mechanical Philosophy,' " 9–10. Thomas Young made the laws of motion mathematical rather than experimental, consequences of his definitions of motion and space. He recognized that the laws were in agreement with experiment, but they were too abstract and simple to be directly proved by experiment; no body was free from extraneous forces, and to take them into account would be to assume what was to be proved. Young acknowledged that his view of the laws of motion was unusual. Thomas Young, *A Course of Lectures on Natural Philosophy and the Mechanical Arts,* 2 vols. (London, 1807) 1:vii, 21–23.

44. Atwood, *Treatise,* i, 2, 357–60.

45. Wolf, *A History of Science, Technology, and Philosophy in the 18th Century* 1: 61–62.

46. Desaguliers, *Experimental Philosophy* 2:72.

47. Hankins, *Science and the Enlightenment,* 30.

48. Erwin N. Hiebert, *Historical Roots of the Principle of Energy Conservation* (Madison: State Historical Society of Wisconsin, 1962), 80–93. Peter Heimann, " 'Geometry and Nature': Leibniz and Johann Bernoulli's Theory of Motion," *Centaurus* 21 (1977): 1–26.

49. Maclaurin, *Fluxions,* 2:427, 432, 439, 452. [Colin Maclaurin], "An Account of a Book Intitled, *A Treatise of Fluxions, in Two Books,* by Colin McLaurin . . . ," *PT* 42 (1743): 325–63, on 353.

50. Heimann, " 'Geometry and Nature,' " 2–3, 18–19. Thomas L. Hankins, "Eighteenth-Century Attempts to Resolve the *Vis viva* Controversy," *Isis* 56 (1965): 281–97, on 282.

51. Maclaurin, *Fluxions* 2:427, 433–34.

52. Parkinson, *Mechanics,* 67–72.

53. Atwood, *Treatise,* 238, 240, 370, 377. In the year that this book was published, Atwood was appointed to a post in the Treasury by William Pitt. Until then, he had been a fellow in Trinity College, Cambridge. Eric M. Cole, "Atwood, George," *Dictionary of Scientific Biography,* ed. C. C. Gillispie, vol. 1 (New York: Charles Scribner's Sons, 1970), 326–27.

54. Desaguliers, *Experimental Philosophy* 2:v–vi.

55. Maclaurin, *Fluxions* 2:453.

56. Ludlam, *Mathematical Essays,* 60.

57. Michell, "On the Means of Discovering the Distance, Magnitude, &c. of the Fixed Stars," 328–43. J. Eisenstaedt, "De l'influence de la gravitation sur la propagation de la lumière en théorie newtonienne. L'archéologie des trous noirs," *Archive for History of Exact Sciences* 42 (1991): 325–86, on 328–43.

58. A. Wolf, *A History of Science, Technology and Philosophy in the 16th and 17th Centuries*, 2d ed., ed. D. McKie, vol. 1 (New York: Harper and Brothers, 1959), 233–34.

Chapter 5

Some sections of this chapter are taken from part 3, chapter 5, "Mercury," in Jungnickel and McCormmach, Cavendish, *The Experimental Life*, 393–423.

1. This and other facts about Cavendish's life are given in Russell McCormmach, "Cavendish, Henry," *Dictionary of Scientific Biography*, ed. C. C. Gillispie, vol. 3 (New York: Charles Scribner's Sons, 1971), 155–59.

2. Schofield, *Mechanism and Materialism*, 254.

3. Richard Kirwan to Sir Joseph Banks, 10 Jan. 1789, copy, Natural History Museum, Botany Library, DTC, 6:122–24.

4. Jungnickel and McCormmach, *Cavendish, the Experimental Life*, 335–54.

5. Quote from a contemporary of Cavendish's, in George Wilson, *The Life of the Honourable Henry Cavendish . . .* (London, 1851), 181–82.

6. 19 Feb. 1807, Charles Blagden Diary, Royal Society, 5:39.

7. Charles Blagden to William Cullen, 17 June 1784, draft, Charles Blagden Letterbook, Beinecke Rare Book and Manuscript Library, Yale University, Osborn Shelves fc15.

8. Truesdell, *The Rational Mechanics of Flexible or Elastic Bodies 1638–1788*; "A Program toward Rediscovering the Rational Mechanics of the Age of Reason," *Archive for History of Exact Sciences* 1 (1960): 1–36, on 17. Giulio Maltese, "Toward the Rise of the Modern Science of Motion: The Transition from Synthetical to Analytical Mechanics," *History of Physics in Europe in the 19th and 20th Centuries*, Italian Physical Society Conference Proceedings, vol. 42, ed. F. Bevilacqua (Bologna: Italian Physical Society, 1993), 51–67. Greenberg, "Mathematical Physics in Eighteenth-Century France," 77.

9. Henry Cavendish, "Plan of a Treatise on Mechanicks," Cavendish Mss VIb, 45: 1.

10. C. Truesdell, "Reactions of Late Baroque Mechanics to Success, Conjecture, Error, and Failure in Newton's *Principia*," *The* Annus Mirabilis *of Sir Isaac Newton 1666–1966*, ed. R. Palter (Cambridge, MA: MIT Press, 1970), 192–232, on 204.

11. Cavendish, "Plan," 13, 16.

12. Robison, *Mechanical Philosophy*, 1:81.

13. Henry Cavendish, "Resistance of Air to Projectiles in Connexion with Newton's Principia," "Investigation of the Loss of Force Owing to the Inertia of the Powder," "To Find the Center of Friction," Cavendish Mss VIb 14, 36, 37. The titles given to Cavendish's mechanical manuscripts in this and the following notes sometimes differ slightly from his; sometimes he gives none.

14. Henry Cavendish, "Series of Experiments on the Flow of Water through a Glass Tube," Cavendish Mss VIb 27.

15. Henry Cavendish, "Concerning the Spinning of Tops by M[r] Mitchell," Cavendish Mss VI(b), 18. In a footnote to this paper, Cavendish wrote: "This point (on which the analysis of the top's motion rests) is what is known by the name of the center of percussion. How I came not to take notice of it I do not know." Truesdell, "Program," 9.

16. Henry Cavendish, "On the Motion of Water Running Out from a Hole in the Bottom of a Vessel," Cavendish Mss VI(b), 26. Truesdell, "Program," 19–20.

17. Henry Cavendish, "Figure of Earth," Cavendish Mss VIII, 15.

18. Henry Cavendish, "Precession of Equinoxes," Cavendish Mss VIII, 9.

19. Henry Cavendish, "On the Motion of Pendulums Whose Centers of Suspension Move," "Pendulum, Records of Swing," "Pendulum, Experiments," "Concerning the Vibrations of Pendulums Suspended from the Same Horizontal Bar," "On Centrifugal Pendulum," "On Error of Pendulum Beating Dead," "Center of Oscillation," Cavendish Mss VIb 17, 19, 20, 24, 25, 29, 40.

20. Henry Cavendish, "Experiments on Twisting of Glass Tubes and Springing of Wires," "Experiments on Twisting of Wire Tried by the Time of a Vibration," "Concerning Springs," "Vibrations of Straight Uniform Spring Left to Itself," Cavendish Mss VIb 21, 22, 28, 41. Truesdell, "Program," 9–10, 19–20.

21. Truesdell, "Program," 9.

22. Henry Cavendish, "Concerning Waves," "On the Motion of Sounds," Cavendish Mss VI(b), 23, 35.

23. Henry Cavendish, "On the Diminution of the Diurnal Motion of the Earth in Consequence of the Tides," Cavendish Mss VIII 10; printed in part in *Scientific Papers* 2:437.

24. Wolf, *History of Science, Technology and Philosophy in the 18th Century* 1:71–72. Hans Straub, "Bernouilli, Daniel," *Dictionary of Scientific Biography*, ed. C. C. Gillispie, vol. 2 (New York: Charles Scribner's Sons, 1970), 36–46, on 43.

25. Daniel Bernoulli, *Hydrodynamica, sive de viribus et motibus fluidorum commentarii* (Strasburg, 1738); *Hydrodynamics by Daniel Bernoulli and Hydraulics by Johann Bernoulli*, 3, 16.

26. Hiebert, *Historical Roots*, 60.

27. This is "a not strictly precise nor full claim to the conservation of mechanical energy." C. Truesdell, *Rational Fluid Mechanics, 1687–1765*, Editor's Introduction to *Euleri Opera omnia*, ser. 2, vol. 12 (Zurich: Orell Füssli, 1954), ix–cxxv, on xxiv.

28. Bernoulli, *Hydrodynamics*, 12–13, 15. Erwin N. Hiebert, Comments, *Critical Problems in the History of Science*, ed. M. Clagett (Madison: University of Wisconsin Press, 1959), 395–96.

29. Daniel Bernoulli, "Rémarques sur le principe de la conservation des forces

vives pris dans un sens général," *Mémoires de l'Académie Royale des Sciences et Belles Lettres de Berlin,* 4 1748 (1750): 356–64; in Daniel Bernoulli, *Die Werke von Daniel Bernoulli,* vol. 3: *Mechanik,* ed. D. Speiser, A. de Baenst-Vandenbroucke, J. L. Pietenpol, and P. Radelet-de Grave (Basel: Birkhäuser, 1987), 197–206. Wolf, *History of Science, Technology and Philosophy in the 18th Century* 1:63.

30. Heimann and McGuire, "Cavendish and the *Vis Viva* Controversy."

31. Henry Cavendish, "Remarks on the Theory of Motion," *Scientific Papers* 2: 415–30.

32. John Robison, the editor of Black's lectures, and like Black a disbeliever in the mechanical theory of heat, found it "amusing" that "mechanical explainers" explained chemical phenomena such as the heat of chemical solutions by "violent knockings of the conflicting particles in the act of solution . . . yet the same act of solution is accompanied with intense cold, when salt and snow are mixed." Footnote in Black, *Lectures* 1:121.

33. Black, *Lectures* 1:80.

34. Irvine, *Essays,* 23.

35. W. H. Wollaston, "The Bakerian Lecture on the Force of Percussion," *PT* 96 (1806): 13–22, on 16. According to Wollaston, Bernoulli first, and then Smeaton, called vis viva "mechanic force." Because Newton did not treat this measure of force, he gave it no definition.

36. Joseph Larmor's editorial note in Cavendish, *Scientific Papers* 2:428.

37. Henry Cavendish, "Watts Fire Place for Burning Smoke," Cavendish Mss Misc.; "Computations & Observations in Journey 1785," Cavendish Mss X(a) 4:36.

38. Cavendish, "Computations & Observations in Journey 1785."

39. Charles Blagden to Sir Joseph Banks, 23 Oct. 1785, Banks Correspondence, Kew, 1:212.

40. Michell, "Conjectures Concerning the Cause, and Observations upon the Phaenomena of Earthquakes," 569.

41. Josiah Wedgwood, "An Attempt to Compare and Connect the Thermometer for Strong Fire, Described in Vol. LXXII of the Philosophical Transactions with the Common Mercurial One," *PT* 74 (1784): 358–84, on 370–71, 381; "Additional Observations on Making a Thermometer for Measuring the Higher Degrees of Heat," *PT* 76 (1786): 390–408.

42. Benjamin Thompson, "An Inquiry Concerning the Source of the Heat Which Is Excited by Friction," *PT* 88 (1798): 80–102; in *Complete Works* 1:491.

43. Black, *Lectures* 1:11–12.

44. Cleghorn, "*De Igne,*" 13–15.

45. Henry Cavendish Mss III(a), 9, untitled bundle.

46. Cavendish Mss III(a), 9:114. His reference is to George Martine's *Essays Medical and Philosophical* (London, 1740).

47. Henry Cavendish, "Experiments to Show That Bodies in Changing from a

Solid State . . . ," *Scientific Papers* 2:348. His source was probably J. J. d'Ortous de Mairan's *Dissertation sur la Glace, ou Explication Physique de la Formation de la Glace, et de ses Divers Phénomènes* (Paris, 1749).

48. Darwin, "Remarks on the Opinion of Henry Eeles, Esq," 245.

49. Benjamin Franklin to Ebenezer Kinnersley, 20 Feb. 1762, in Benjamin Franklin, *Benjamin Franklin's Experiments. A New Edition of Franklin's Experiments and Observations on Electricity*, ed. I. B. Cohen (Cambridge, MA: Harvard University Press, 1941), 360. Franklin was in London in 1762 as the agent of the Pennsylvania Assembly.

50. Boerhaave's *A New Method of Chemistry; Including the Theory and Practice of That Art* . . . (London, 1727) is listed in Christopher Wordsworth, *Scholae Academicae: Some Account of the Studies at the English Universities in the Eighteenth Century* (Cambridge, 1877), 79.

51. Joseph Black, *Notes from Doctor Black's Lectures on Chemistry 1767/8*, ed. D. McKie (Cheshire: Imperial Chemical Industries, 1966), 8. Guerlac, "Black, Joseph," 177–78. Brook Taylor, "An Account of an Experiment, Made to Ascertain the Proportion of the Expansion of the Liquor in the Thermometer, with Regard to Degrees of Heat," *PT* 32 (1723): 291. Wolf, *History of Science, Technology, and Philosophy in the 18th Century* 1:189–90. George Wilson, *Cavendish*, 447.

52. Henry Cavendish, "Experiments on Heat," *Scientific Papers* 2:327–51, on 327.

53. Cavendish Mss III(a) 9:76.

54. Cavendish, "Experiments on Heat," 332.

55. Ibid., 340.

56. Cavendish Mss III(a), 9:22, 27–38.

57. Henry Cavendish, "Concerning Heat & Cold Produced by Hardening & Melting of Spermaceti," Cavendish Mss III(a), 9:31. Cavendish measured the latent heat of spermaceti several times, obtaining roughly consistent results; he found that the hardening of spermaceti was sufficient to raise an equal weight of water by about sixty-four to seventy-five degrees, and the cold generated by the melting of it fell in that range. Using a different experimental arrangement, Cavendish returned to spermaceti, getting higher values this time. Ibid., 78–81. His experiments on the change of state of water between solid, liquid, and vapor gave much more consistent results, and these he emphasized in his intended paper. He only briefly mentioned there that he had found the same kind of result upon hardening and melting spermaceti. "Experiments to Show That Bodies in Changing from a Solid State to a Fluid State Produce Cold and in Changing from a Fluid to a Solid State Produce Heat," *Scientific Papers* 2:348–50.

58. Cavendish Mss III(a), 9:42–47. Cavendish wrote up this experiment in his paper "Experiments on Heat," 345–46.

59. Cavendish, "Experiments on Heat," 343.

60. Cavendish nowhere mentioned Irvine by name, but he wrote two short memoranda on principal points of Irvine's theory. One of them simply states a propo-

sition from Irvine's theory, which says that any heat that appears in bodies depends entirely on their heat capacities and changes in these heats; the proposition is expressed in the language of the material theory of heat, with heat said to be "absorbed" by bodies and "united" to them. This proposition is Irvine's explanation of latent heats. Henry Cavendish, "That All the Heat Which Appears in Bodies . . . ," Cavendish Mss Misc.

61. The second memorandum is an experimental "compleat proof" that the absolute heat in bodies is not proportional to their specific heats, as Irvine's theory required. Henry Cavendish, "A Compleat Proof . . . ," Cavendish Mss Misc.

62. Cavendish Mss III(a) 9:39–40.

63. Henry Cavendish, "Hypothesis," Cavendish Mss Misc.

64. J. A. Deluc to Henry Cavendish, 19 Feb. 1777, Cavendish Mss, New Correspondence; in Jungnickel and McCormmach, *Cavendish, the Experimental Life*, 546–48.

65. Henry Cavendish, "Theory of Boiling," *Scientific Papers* 2:354–62, on 354, 358, 361–62.

66. George Wilson, *Cavendish*, 446. In agreement with Wilson: McKie and Heathcote, *Discovery*, 52.

67. In 1772, the Swedish physicist Johan Carl Wilcke discovered latent heat independently of Black; unlike Black, he published his finding. McKie and Heathcote, *Discovery*, 54–121.

68. This "circumstance" is "that all, or almost all, bodies by changing from a fluid to a solid state, or from the state of an elastic to that of an unelastic fluid, generate heat; and that cold is produced by the contrary process." Henry Cavendish, "Observations on Mr. Hutchins's Experiments for Determining the Degree of Cold at Which Quicksilver Freezes," *PT* 73 (1783): 303–28; in *Scientific Papers* 2:145–60, on 150.

69. Cavendish, "Observations on Mr. Hutchins's Experiments," 146.

70. Henry Cavendish, "An Account of Experiments Made by Mr. John McNab, at Henly House, Hudson's Bay, Relating to Freezing Mixtures," *PT* 76 (1786): 241–72; in *Scientific Papers* 2:195–213, on 195.

71. Henry Cavendish, "An Account of Experiments Made by Mr. John McNab, at Albany Fort, Hudson's Bay, Relative to the Freezing of Nitrous and Vitriolic Acids," *PT* 78 (1788): 166–81; in *Scientific Papers* 2: 214–23.

Chapter 6

Some sections of this chapter are taken from part 3, chapter 5, "Mercury," in Junknickel and McCormmach, *Cavendish, The Experimental Life*, 393–423.

1. Nicholson, *Natural Philosophy* 2: 122.

2. Cavendish, "Observations on Mr. Hutchin's Experiments," in *Scientific Papers* 2:145–60, quotation on pp. 150–51.

3. Cavendish, "Experiments on Air," 161–181, on 173–74.

4. James Hutton, *Philosophy of Light, Heat, and Fire*, 7–9.

5. Henry Cavendish, "Remarks on the Theory of Motion," *Scientific Papers* 2:415–30, corollary 2 on pp. 425–26; "Experiments on Heat," ibid., 327–51, on 351.

6. Russell McCormmach, "Henry Cavendish: A Study of Rational Empiricism in Eighteenth-Century Natural Philosophy," *Isis* 60 (1969): 293–306.

7. Maclaurin, *Account*, 12, 18–19; Locke, *Essay*, 332.

8. Desaguliers, *Experimental Philosophy* 1:preface.

9. Ravetz, "The Representation of Physical Quantities," 7–20, on 7, 15.

10. Charles Walmesley, "Two Essays Addressed to the Rev. James Bradley, D.D.," PT 49 (1755): 700–59, on 703.

11. Hankins, "Newton's 'Mathematical Way," 90. August Claude Boissonnade and Victor N. Vagliente, "Translator's Introduction," in J. L. Lagrange, *Analytical Mechanics*, trans. and ed. A. C. Boissonnade and V. N. Vagliente (Dordrecht, Kluwer Academic Publishers, 1997), xi–xli, on xxxiv–xli.

12. Issac Milner, "Essay," Cambridge University Library, Ee 5.35, pp. 50, 94. L. J. M. Coleby, "Isaac Milner and the Jacksonian Chair of Natural Philosophy," *Annals of Science* 10 (1954): 234–57, on 250–51.

13. Milner; "Essay," 31. Smith, *Harmonics*, xv.

14. Henry Cavendish, "Heat," Manuscript Division, Pre-Confederation Archives, Public Archives of Canada, Ottawa, M G 23, L 6, pp. 28–29; foul copy, pp. 17–18. Boscovich, *Theory of Natural Philosophy*, 21–23. Hankins, "Eighteenth-Century Attempts to Resolve the *Vis viva* Controversy," 291–97. Michell arrived independently at views similar to Boscovich's, and Cavendish may have done so, too. It has been pointed out that there was a British tradition paralleling Boscovich's views. Cantor, *Optics after Newton*, 71–72; Schofield, *Mechanism and Materialism*, 236–49; Heimann and McGuire, "Newtonian Forces and Lockean Powers."

15. Newton, *Principia* 2:547; *Opticks*, 348–54.

16. Cavendish, *Electrical Researches*, 103, 410.

17. John C. Greene, *Science, Ideology, and World View: Essays in the History of Evolutionary Ideas* (Berkeley: University of California Press, 1981), 11–16.

18. Newton, *Principia* 1:19.

19. Cavendish, "Remarks," *Scientific Papers* 2:423. Newton, Book 1, Proposition 40, *Principia* 1:128.

20. Daniel Bernoulli, "Rémarques sur le principe de la conservation des forces vives pris dans un sens général," 205; editors' commentary, 99.

21. Cavendish, "Heat," 14–16.

Chapter 7

Some sections of this chapter are taken, from part 3, chapter 5, "Mercury," in Jungkenickel and McCormmach, Cavendish, *The Experimental Life*, 393–423.

1. Cantor, *Optics after Newton*, 57. Priestley, *Vision, Light and Colours* 1:387–89. S. G. Brush and C. W. F. Everitt, "Maxwell, Osborne Reynolds, and the Radiometer," *Historical Studies in the Physical Sciences* 1 (1969): 103–25.

2. Henry Cavendish, "Heat," Manuscript Division, Pre-Confederation Archives, Public Archives of Canada, Ottawa, M G 23, L 6, p. 22.

3. Richard Watson, "An Account of an Experiment Made with a Thermometer, Whose Bulb Was Painted Black, and Exposed to the Direct Rays of the Sun," *PT* 63 (1773): 40–41.

4. Tiberius Cavallo, "Thermometrical Experiments and Observations," *PT* 70 (1780): 587–99, on 591–94.

5. Fordyce, "An Account of an Experiment on Heat."

6. Cavendish, "Heat," 22.

7. Ibid., foul copy, 12.

8. Hiebert, *Historical Roots*, 102.

9. Heinz Otto Sibum, "Reworking the Mechanical Value of Heat: Instruments of Precision and Gestures of Accuracy in Early Victorian England," *Studies in the History and Philosophy of Science* 26 (1995): 73–106, on 73–74, 104–5.

10. Thomas Young, *Natural Philosophy and the Mechanical Arts*, 1:655.

11. Irvine, *Essays*, 171.

12. Thompson, "An Inquiry Concerning the Source of Heat Which Is Excited by Friction." Robison, *Mechanical Philosophy* 1:611.

13. William Nicholson, *The First Principles of Chemistry* (London, 1790), 6.

14. Cavendish, "Heat," 40, 43.

15. Ibid., foul copy, 15.

16. Irvine, *Essays*, 11–12.

17. Cavendish, "Heat," 41–42.

18. John Canton, "An Attempt to Account for the Regular Diurnal Variation of the Horizontal Magnetic Needle; and Also for Its Irregular Variation at the Time of an Aurora Borealis," *PT* 51 (1759): 398–445, on 400.

19. John Michell, *A Treatise of Artificial Magnets; in Which Is Shewn an Easy and Expeditious Method of Making Them, Superior to the Best Natural Ones; and Also, a Way of Improving the Natural Ones, and of Changing or Converting Their Poles. Directions Are Likewise Given for Making the Mariner's Needles in the Best Form, and for Touching Them Most Advantageously* (Cambridge, 1750), 24.

20. Benjamin Wilson, "Experiments on the Tourmalin," *PT* 51 (1759): 308–39, on 335–37.

21. C. Truesdell, *Rational Fluid Mechanics, 1687–1765*, cvii–viii.

22. Henry Cavendish, "Cold Produced by Rarefaction of Air," "Heat and Cold by Exhausting and Condensing of Air," *Scientific Papers* 2:384, 385–89.

23. Daniel Bernoulli, *Hydrodynamics* 1:165–71. Henry Guerlac, "Chemistry as a Branch of Physics: Laplace's Collaboration with Lavoisier," *Historical Studies in the Physical Sciences* 7 (1976): 193–276, on 246–49.

24. John Playfair, *The Works of John Playfair*, 4 vols., ed. J. G. Playfair (Edinburgh, 1822) 1:lxxxiii–lxxxiv.

25. Charles Blagden to A. L. Lavoisier, 15 Sept. 1783, draft, Blagden Letterbook, Beinecke Rare Book and Manuscript Library, Yale University, Osborn Shelves fc15.

26. A. L. Lavoisier and P. S. Laplace, *Memoir on Heat. Read to the Royal Academy of Sciences, 28 June 1783 by Messrs. Lavoisier and De La Place*, trans. H. Guerlac (New York: Neale Watson Academic Publications, 1982), 4–6. Guerlac, "Chemistry as a Branch of Physics," 244–48. David V. Fenby, "Chemical Reactivity and Heat in the Eighteenth Century," *Philosophy and Science in the Scottish Enlightenment*, ed. P. Jones (Edinburgh: John Donald, 1988), 67–86, on 81.

27. Charles Blagden to Antoine Laurent Lavoisier, draft, 15 Sept. 1783, Blagden Letterbook, Yale University Library, Osborn Shelves fc15.

28. Fordyce, "An Account of Some Experiments on the Loss of Weight in Bodies on Being Melted or Heated."

29. Charles Blagden to P. S. Laplace, 5 Apr. 1785; Charles Blagden to C. L. Berthollet, 28 June, 1785, Blagden Letterbook, Yale. Charles Blagden to Henry Cavendish, n.d., [after 2 June 1785], Jungnickel and McCormmach, *Cavendish, the Experimental Life*, 608–9.

30. John Roebuck, "Experiments on Ignited Bodies," *PT* 66 (1776): 509–12. These experiments, witnessed by Cavendish, showed an increase of weight in iron and silver upon cooling.

31. Cavendish, sketch of paper, "Heat."

32. Young, *Natural Philosophy and the Mechanical Arts* 1:457.

33. Nicholson, *Natural Philosophy* 1:134; *First Principles of Chemistry*, 6.

34. Bryan Higgins, *Experiments and Observations Relating to Acetous Acid, Fixable Air, Dense Inflammable Air, Oils and Fuel; the Matter of Fire and Light . . .* (London, 1786), 301–2.

35. David Gregory, *The Elements of Astronomy, Physical and Geometrical*, 2 vols. (London, 1715) 1:ii.

36. Cavendish, "Heat," 36.

37. Cavendish was right in thinking that there is no reason why bodies can only expand with heat and contract with cold. He evidently did not know that between four degrees Celcius and the freezing point, water expands upon cooling. See also comment on this point in note in appendix.

38. John Leslie, *An Experimental Inquiry into the Nature, and Propagation of Heat* (London, 1804), 140–41.

39. Adams, *Natural and Experimental Philosophy*, 1:207.

40. Young, *Natural Philosophy and the Mechanical Arts* 1:656.

41. Henry Cavendish to John Michell, 27 May 1783, Jungnickel and McCormmach, *Cavendish, the Experimental Life*, 567–69.

42. Cavendish, "Heat," 42.

43. William Hillary, *The Nature, Properties, and Laws of Motion of Fire Discovered and Demonstrated by Observations and Experiments* (London, 1760), preface.

44. Robison, *Mechanical Philosophy* 1:265–66, 4:1–2.

45. James Hutton, *Philosophy of Light, Heat, and Fire*, xi.

46. Wilson, *Cavendish*, 185–86.

47. Hamilton, *Philosophical Essays*, 36.

48. These are examples. Stefan L. Wolff, "Origins of Theoretical Physics in Germany in 19th Century," *I Beni Culturali Scientifici nella Storia e Didattica. Atti del Convego del 14–15 dicembre 1990* (Pavia: Università degli Studi di Pavia, 1990), 162–76. S. D'Agostino, "A Consideration of the Rise of Theoretical Physics in Europe and of Its Interaction with the Philosophical Tradition," *History of Physics in Europe in the 19th and 20th Centuries*, ed. F. Bevilacqua, Conference Proceedings of the Italian Physical Society, vol. 42 (Bologna: Italian Physical Society, 1992), 5–28. Arcangelo Rossi, "Kantism, Phenomenalism, Reductionism and the Emergence of Theoretical Physics in the 19th Century," ibid., 279–85. Christa Jungnickel and Russell McCormmach, *Intellectual Mastery of Nature: Theoretical Physics from Ohm to Einstein*, vol. 1: *The Torch of Mathematics, 1800–1870* (Chicago: Chicago University Press, 1986).

49. Peter Harman, *The Natural Philosophy of James Clerk Maxwell* (Cambridge: Cambridge University Press, 1998), 3.

Appendix: Editorial Note

1. The original manuscript of "Heat" is located under the reference M G 23, L 6 in the Manuscript Division, Pre-Confederation Archives, Public Archives of Canada, Ottawa.

2. At auction sales, "Heat" was assigned first to the decade 1795–1805 and then to around 1780; the truth probably lies somewhere between. Cavendish certainly wrote this paper after "Remarks on the Theory of Motion," which mentions only some of the phenomena discussed in "Heat." *Scientific Papers* 2:415–30. Also, in "Remarks" Cavendish regarded the cold produced by chemical mixtures as a difficulty for the theory, whereas in "Heat" he no longer did. Most important for this comparison is that in "Heat" Cavendish drew on his knowledge of specific and latent heats, developing the mechanical theory accordingly, whereas in "Remarks" he did not mention them. The connection between "Heat" and Cavendish's experiments on specific and latent heats is direct; for example, the numbered paragraph 7 on p. 16 of "Heat,"

concerning the heats of chemical mixtures, states in general terms the conclusion on p. 39 of the experimental notes on heat, Cavendish Mss III(a), 9. Christie's sales catalogue assigned the first dating primarily on the basis of the watermarks of the paper, in which the name J. Cripps alternates with Britannia in a crowned circle. The assumption was that the earliest recorded mark of James Cripps was in 1792. Although Cavendish did use the J. Cripps stationery several times after that year, he also used it earlier, in the 1780s (the earliest appearance being manuscript pages A3 through A5 of "Experiments on Air," Cavendish Mss II, 10, published in the *Philosophical Transactions* in 1785). "Heat" reappeared at Dawsons of Pall Mall, which noted that James Cripps, father and son, made paper from 1753 to 1803. Based on references to other authors in the manuscript, a new dating was proposed, around 1780.

3. Cavendish, "Heat," 23. Cavendish's source was undoubtedly the experiments on "heat rays" and light using polished metal and glass, discussed in Carl Wilhelm Scheele, *Chemical Observations and Experiments on Air and Fire. . . .*, 1777, trans. J. R. Forster, with notes by Richard Kirwan (London, 1780), 72–74, 92–98.

4. Cavendish, "Heat," 23. Here Cavendish's source was no doubt H. B. de Saussure's account of the experiments he did with M. A. Pictet on the reflection of "obscure heat" emitted by hot, but not red-hot, bodies, reported in *Voyages dans les Alpes. . . .*, vol. 2 (Neuchâtel, 1786), 354–55.

5. For example, Pierre Prevost's experiments on heat rays and Count Rumford's on the mechanical production of heat, belonging to the 1790s, would have been relevant to Cavendish's argument, as would William Herschel's experiments on radiant heat from 1800.

6. Bennet, "A New Suspension of the Magnetic Needle"; *New Experiments on Electricity. . . .* (Derby, 1789).

Appendix: [Heat]

1. This is the dot notation of Newtonian fluxional calculus, only here it stands for the increment of B in a short time, not the fluxion of B. This was commonly done then, whereas Newton would have denoted the increment by $\dot{B}o$.

2. Cavendish here begins his derivation of the principle of conservation of energy. This principle, as we know, states that the quantity of energy of a material system cannot be changed by any interactions of the parts of the system; the form the energy takes, however, can change, the reason for the power and generality of the principle. Heat is one such form, Cavendish argues in this paper. (His reasoning will seem unfamiliar to a reader today, who is not helped by Cavendish's brevity nor by his multiple use of the same symbols; B, for instance, alternately stands for particle B, its vis viva, and its position.)

His derivation follows from the nature of the forces he assumes, attractions and repulsions that depend only on the separations of particles. Granted this assumption, if the vis viva, the energy of motion, or (twice) the kinetic energy, of the particles of a system is written in modern notation as T, and their energy of position, or potential energy, as V, he proves that the total energy T + V does not change over time: $d/dt\ (T + V) = 0$.

In his earlier derivation of the principle, in "Remarks," he drew explicitly on Proposition 40, Book I of the *Principia*; it helps in following the new derivation, too. An implication is that a particle attracted to a center acquires the same increment of velocity, or vis viva, in falling through the same space regardless of the path it takes or the velocity with which it enters the space. Further, the increment of vis viva of a particle acted upon by the united forces of several particles is equal to the sum of the increments of vis viva of the particle generated by the other particles acting singly.

Owing to their mutual forces, the particles of a material system change their configuration and motion from one instant to the next. Cavendish analyzes this change for a system consisting of four particles—B, D, E, and F—then generalizes his result. Denoting by $\dot{B}$ the increase of vis viva of particle B when the configuration of the system changes ever so slightly during a vanishingly short time, he resolves the increase into contributions arising from the actions of each of the other three particles: The partial increase of vis viva of particle B, in the direction in which it is moving, owing to the action of particle D is $\overline{bd}^{\cdot}$, that owing to the action of particle E is $\overline{be}^{\cdot}$, and that owing to the action of particle F is $\overline{bf}^{\cdot}$: Thus, $\dot{B} = \overline{bd}^{\cdot} + \overline{be}^{\cdot} + \overline{bf}^{\cdot}$. By Newton's theorem, particle B would have acquired the same partial increment of vis viva from D if it had moved on a straight line to or from D to a position equally distant from D, independently of the actions of the other particles on B. Cavendish rewrites $\dot{B}$ accordingly, replacing the original b with Greek letter beta: $\dot{B} = \overline{\beta d}^{\cdot} + \overline{\beta e}^{\cdot} + \overline{\beta f}^{\cdot}$. He applies the same reasoning to the other pairs of particles. The rest is arithmetic and grouping of terms. The final equation reads 0 on one side, and on the other side the difference between the sum of the actual increments of vis viva of the particles produced by the combined action of all of the other particles, that is, of the increment of T, and the sum of the computed increments of vis viva that the particles would receive from all of the other particles if they were to act singly, along their axis, that is, of the increment of V. The factor one half is needed because as D attracts B, B attracts D, etc. The result states that if T increases, V must decrease by an equal quantity, and vice versa: If the actual vis viva of the system increases, the potential vis viva of the system decreases, and vice versa. That is, the total vis viva of the system is not altered by the interactions of the particles with one another. Total vis viva is conserved.

3. John Michell, 1724–93, English minister and natural philosopher. His experiment is discussed on p. 73.

4. Joseph Priestley, 1733–1804, English minister and natural philosopher.

5. Cavendish gives the velocity of light in inches per second. 12,000,000,000 inches per second is equivalent to 189,000 miles per second, close to our value of 186,000. It is convenient for Cavendish's rough calculation that the reciprocal of the number of feet in a mile, a human convention, is, with proper placing of the decimal, so nearly the same as a universal constant, the speed of light in the vacuum.

6. Carl Wilhelm Scheele, 1742–86, Swedish chemist.

7. Horace Bénédict de Saussure, 1740–99, Swiss geologist.

8. R. J. Boscovich, 1711–87, Croatian natural philosopher.

9. Cavendish apparently was unaware that below 4° Celsius water expands with cooling. Not long after, this "singular" property of water became well known. J. A. Deluc had announced that water reaches its maximum density above the freezing temperature, and in 1797 Benjamin Thompson, Count Rumford developed the beneficial implications for all life on earth of this "miraculous" exception to the "general law of nature" that all bodies contract upon cooling. Benjamin Thompson, "On the Propagation of Heat in Fluids," *Complete Works* 1: 239–400, on 308–33. Directing readers to Rumford's discussion, Tiberius Cavallo included the exceptional property in his *Natural or Experimental Philosophy*, 3:35–37.

10. This section follows the conclusion, to be integrated in the revised copy.

11. Henry Cavendish, "An Attempt to Explain Some of the Principal Phaenomena of Electricity, by Means of an Elastic Fluid," *PT* 61 (1771): 584–677. Here he applies his theory to the electric jar. Without a familiarity with his paper of 1771, the reader may have difficulty following his reasoning in "Heat." By making assumptions about the quantity of redundant electric fluid, or charge, set in motion by the discharge of an electric jar, he concludes that the vis viva of the fluid is inadequate to explain the heat generated; he proposes another reason, a change in the latent heat of the discharge wire.

12. This figure and the corresponding incomplete analysis are crossed out, replaced by the next figure and discussion.

13. In this revision, the dot beside the bar over paired letters, used in the first version, is dropped, as unnecessary.

14. The last sentence Cavendish wrote in a small hand having clearly added it later.

BIBLIOGRAPHY OF PUBLISHED WORKS

Abbreviations

PT	*Philosophical Transactions of the Royal Society of London*
TRIA	*Transactions of the Royal Irish Academy*
TRSE	*Transactions of the Royal Society of Edinburgh*

Adams, George. *Lectures on Natural and Experimental Philosophy.* 5 vols. London, 1794.

Akenside, Mark. "Observations on the Origin and Use of the Lymphatic Vessels of Animals. . . ." *PT* 50 (1757): 322–28.

Atwood, George. *An Analysis of a Course of Lectures on the Principles of Natural Philosophy, Read in the University of Cambridge. . . .* London, 1784.

———"A General Theory for the Mensuration of the Angeles Subtended by Two Objects, of Which One Is Observed by Rays after Two Reflections from Plane Surfaces, and the Other by Rays Coming Directly to the Spectator's Eye." *PT* 71 (1781): 395–435.

———. "Investigations, Founded on the Theory of Motion, for Determining the Times of Vibration of Watch Balances." *PT* 84 (1794): 119–68.

———. *A Treatise on the Rectilinear Motion and Rotation of Bodies; with a Description of Original Experiments Relative to the Subject.* Cambridge, 1784.

Barnes, Barry, David Bloor, and John Henry. *Scientific Knowledge: A Sociological Analysis.* Chicago: University of Chicago Press, 1996.

Bazerman, Charles. *Shaping Written Knowledge: The Genre of the Experimental Article in Science.* Madison: University of Wisconsin Press, 1988.

Bechler, Zev. "Newton's Ontology of the Force of Inertia." In *The Investigation of Difficult Things: Essays on Newton and the History of the Exact Sciences*, edited by P. M. Harman and A. E. Shapiro, 287–304. Cambridge: Cambridge University Press, 1992.

Bennet, Abraham. *New Experiments on Electricity. . . .* Derby, 1789.

———. "A New Suspension of the Magnetic Needle, Intended for the Discovery of Minute Quantities of Magnetic Attraction: Also an Air Vane of Great Sensibility; with New Experiments on the Magnetism of Iron Filings and Brass." *PT* 82 (1792): 81–98.

Bentham, Jeremy. *Bentham's Theory of Fictions*. Edited by C. K. Ogden. New York: Harcourt, Brace, 1932.

Berkeley, George. *A Treatise Concerning the Principles of Human Knowledge*. Edited by C. P. Krauth. Philadelphia: J. B. Lippincott, 1881.

Bernoulli, Daniel. "Rémarques sur le principe de la conservation des forces vives pris dans un sens général." *Mémoires de l'Académie Royale des Sciences et Belles Lettres de Berlin* 4 1748 (1750): 356–64. In Bernoulli, *Werke* 3: 197–206. *Die Werke von Daniel Bernoulli*. Vol. 3: *Mechanik*. Edited by D. Speiser, A. de Baenst-Vandenbroucke, J. L. Pietenpol, and P. Radelet-de Grave. Basel: Birkhäuser, 1987.

Bernoulli, Daniel, and Johann Bernoulli. *Hydrodynamics by Daniel Bernoulli and Hydraulics by Johann Bernoulli*. Translated by T. Carmody and H. Kobus. New York: Dover, 1968.

Black, Joseph. *Lectures on the Elements of Chemistry*. Edited by J. Robison. 3 vols. Philadelphia, 1806–7.

———. *Notes from Doctor Black's Lectures on Chemistry 1767/8*. Edited by D. McKie. Cheshire: Imperial Chemical Industries, 1966.

Black, Joseph, and William Cullen. *Experiments upon Magnesia Alba, Quick-Lime, and Other Alkaline Substances; by Joseph Black. To Which Is Annexed, An Essay on the Cold Produced by Evaporating Fluids, and of Some Other Means of Producing Cold; by William Cullen*. Edinbugh, 1777.

Blackstone, William. *Commentaries on the Laws of England*. 13th edition. London, 1800.

Blagden Charles. "An Account of Some Late Fiery Meteors: With Observations." *PT* 74 (1784): 201–32.

———. Obituary of Henry Cavendish. *Gentleman's Magazine*, March 1810, 292.

Blair, Robert. "Experiments and Observations on the Unequal Refrangibility of Light." *TRSE* 3:2 (1791): 3–76.

Boerhaave, Herman. *A New Method of Chemistry; Including the Theory and Practice of That Art. . . .* London, 1727.

Boissonnade, Auguste Claude, and Victor N. Vagliente. "Translator's Preface." In J. L. Lagrange, *Analytical Mechanics*, translated and edited by A. C. Boissonnade and V. N. Vagliente, xi–xli. Dordrecht: Kluwer Academic Publishers, 1997.

Boscovich, Roger Joseph. *A Theory of Natural Philosophy*. Translated by J. M. Child from the 2d edition of 1763. Cambridge, MA: MIT Press, 1966.

Brush, S. G., and C. W. F. Everitt. "Maxwell, Osborne Reynolds, and the Radiometer." *Historical Studies in the Physical Sciences* 1 (1969): 105–25.

Buchdahl, Gerd. *The Image of Newton and Locke in the Age of Reason*. London: Sheed and Ward, 1961.

Canton, John. "An Attempt to Account for the Regular Diurnal Variation of the

Horizontal Magnetic Needle; and Also for Its Irregular Variation at the Time of an Aurora Borealis." *PT* 51 (1759): 398–445.

Cantor, G. N. "Anti-Newton." In *Let Newton Be!* edited by J. Fauvel, R. Flood, M. Shorthand, and R. Wilson, 203–22. Oxford: Oxford University Press, 1988.

———. *Optics after Newton: Theories of Light in Britain and Ireland, 1704–1840.* Manchester: Manchester University Press, 1983.

———. "Was Thomas Young a Wave Theorist?" *American Journal of Physics* 52 (1984): 305–8.

Cavallo, Tiberius. *A Complete Treatise of Electricity in Theory and Practice; with Original Experiments.* London, 1777.

———. *The Elements of Natural or Experimental Philosophy.* 4 vols. London, 1803.

———. "Of the Methods of Manifesting the Presence, and Ascertaining the Quality, of Small Quantities of Natural or Artificial Electricity." *PT* 78 (1788): 1–22.

———. "Thermometrical Experiments and Observations." *PT* 70 (1780): 589–99.

———. *A Treatise on Magnetism in Theory and Practice; with Original Experiments.* London, 1787.

Cavendish, Henry. "An Account of Experiments Made by Mr John McNab, at Albany Fort, Hudson's Bay, Relative to the Freezing of Nitrous and Vitriolic Acids." *PT* 78 (1788): 166–81.

———. "An Account of Experiments Made by Mr. John McNab, at Henly House, Hudson's Bay, Relating to Freezing Mixtures." *PT* 76 (1786): 241–72.

———. "An Attempt to Explain Some of the Principal Phaenomena of Electricity, by Means of an Elastic Fluid." *PT* 61 (1771): 584–677.

———. *The Electrical Researches of the Honourable Henry Cavendish.* Edited by J. C. Maxwell. Cambridge: Cambridge University Press, 1879. London: Frank Cass reprint, 1967.

———. "Experiments on Air." *PT* 74 (1784): 119–69.

———. "Experiments to Determine the Density of the Earth." *PT* 88 (1798): 469–526.

———. "Observations on Mr. Hutchins's Experiments for Determining the Degree of Cold at Which Quicksilver Freezes." *PT* 73 (1783): 303–28.

———. "On the Height of the Luminous Arch Which Was Seen on Feb. 23, 1784." *PT* 80 (1790): 101–5.

———. *The Scientific Papers of the Honourable Henry Cavendish, F.R.S.* Edited by E. Thorpe et al. 2 vols. Cambridge: Cambridge University Press, 1921.

Cleghorn, William. "William Cleghorn's *De Igne* (1779)." Edited and translated by D. McKie and N. H. de V. Heathcote. *Annals of Science* 14 (1958): 1–82.

Cline, Barbara Lovett. *Men Who Made a New Physics: Physicists and the Quantum Theory.* New York: New American Library, 1969.

Cohen, I. B. "Commentary by I. Bernard Cohen." In *Scientific Change: Historical Studies in the Intellectual, Social and Technical Conditions for Scientific Discovery and Technical Invention, from Antiquity to the Present*, edited by A. C. Crombie, 466–71. New York: Basic Books, 1961.

———. *Franklin and Newton: An Inquiry into Speculative Newtonian Experimental Science and Franklin's Work in Electricity as an Example Thereof.* Philadelphia: American Philosophical Society, 1956.

———. "Newton's Second Law and the Concept of Force in the *Principia*." In *The* Annus Mirabilis *of Sir Isaac Newton*, edited by R. Palter, 143–85. Cambridge, MA: MIT Press, 1970.

———. "The *Principia*, the Newtonian Style, and the Newtonian Revolution in Science." *Action and Reaction: Proceedings of a Symposium to Commemorate the Tercentenary of Newton's Principia*," edited by P. Theerman and A. F. Seeff, 61–104. Newark: University of Delaware Press, 1993.

Cole, Eric M. "Atwood, George." In *Dictionary of Scientific Biography*, edited by C. C. Gillispie, vol. 1, 326–27. New York: Charles Scribner's Sons, 1970.

Coleby, L. J. M. "Isaac Milner and the Jacksonian Chair of Natural Philosophy." *Annals of Science* 10 (1954): 234–57.

Cotes, Roger. *Hydrostatical and Pneumatical Lectures.* Edited by R. Smith. London, 1738.

Crawford, Adair. *Experiments and Observations on Animal Heat, and the Inflammation of Combustible Bodies. Being an Attempt to Resolve These Phaenomena into a General Law of Nature.* London, 1779.

Crosland, Maurice, and Crosbie Smith. "The Transmission of Physics from France to Britain: 1800–1840." *Historical Studies in the Physical Sciences* 9 (1978): 1–61.

Cunningham, Andrew. "Getting the Game Right: Some Plain Words on the Identity and Invention of Science." *Studies in the History and Philosophy of Science* 19 (1988): 365–89.

———. "How the *Principia* Got Its Name: or, Taking Natural Philosophy Seriously." *History of Science* 29 (1991): 377–92.

D'Agostino, S. "A Consideration of the Rise of Theoretical Physics in Europe and of Its Interaction with the Philosophical Tradition." In *History of Physics in Europe in the 19th and 20th Centuries*, edited by F. Bevilacqua, 5–28. Conference Proceedings of the Italian Physical Society. Vol. 42. Bologna: Italian Physical Society, 1992.

Darwin, Erasmus. "Remarks on the Opinion of Henry Eeles, Esq; Concerning the Ascent of Vapour, Published in the Philosoph. Transact. Vol. XLIX. Part I, P. 124." *PT* 50 (1757): 240–54.

———. *Zoonomia; or The Laws of Organic Life.* 2 vols. London, 1794–96.

Deluc, J. A. *An Elementary Treatise on Geology: Determining Fundamental Points in*

That Science, and Containing an Examination of Some Modern Geological Systems, and Particularly of the Huttonian Theory of the Earth. Translated by H. de la Fite. London, 1809.

Desaguliers, J. T. *A Course of Experimental Philosophy.* 2 vols. London, 1734–44.

———. "Some Things Concerning Electricity." *PT* 41 (1740): 634–37.

———. "Some Thoughts and Conjectures Concerning the Cause of Elasticity." *PT* 41 (1739): 175–85.

———. *A System of Experimental Philosophy.* London, 1719.

Donovan, A. L. *Philosophical Chemistry in the Scottish Enlightenment: The Doctrines and Discoveries of William Cullen and Joseph Black.* Edinburgh: Edinburgh University Press, 1975.

Eeles, Henry. "Concerning the Cause of the Ascent of Vapour and Exhalation, and Those of Winds; and of the General Phaenomena of the Weather and Barometer." *PT* 49 (1755): 124–54.

Einstein, Albert. "Autobiographical Notes." In *Albert Einstein: Philosopher-Scientist,* edited by P. A. Schilpp, vol. 1, 1–94. London: Open Court, 1949.

———. "Foreword." In Galileo, *Dialogues,* vi–xix.

———. "Foreword." In Newton, *Opticks,* lix–lx.

———. "The Fundaments of Theoretical Physics." In *Out of My Later Years,* 95–107.

———. *Ideas and Opinions.* Translated by S. Bargmann. New York: Dell, 1954.

———. "Isaac Newton." In *Out of My Later Years,* 201–4.

———. "The Mechanics of Newton and Their Influence on the Development of Theoretical Physics." In *Ideas and Opinions,* 247–55.

———. *Out of My Later Years.* Totowa, NJ: Littlefield, Adams, 1967.

———. "Principles of Theoretical Physics." In *Ideas and Opinions,* 216–19.

Eisenstaedt, J. "De l'influence de la gravitation sur la propagation de la lumière en théorie newtonienne. L'archéologie des trous noirs." *Archive for History of Exact Sciences* 42 (1991): 325–86.

Ellicott, John. "Several Essays towards Discovering the Laws of Electricity." *PT* 45 (1748): 195–224.

Emerson, William. *The Doctrine of Fluxions: Not Only Explaining the Elements Thereof, but Also Its Application and Use in the Several Parts of Mathematics and Natural Philosophy.* 3d edition. London, 1768.

———. *The Principles of Mechanics. Explaining and Demonstrating the General Laws of Motion, the Laws of Gravity, Motion of Descending Bodies. . . .* 2d edition. London, 1758.

Enfield, William. *Institutes of Natural Philosophy, Theoretical and Experimental. . . .* London, 1785.

English, John C. "John Hutchinson's Critique of Newtonian Heterodoxy." *Church History* 68 (1999): 581–97.

Fenby, David V. "Chemical Reactivity and Heat in the Eighteenth Century." In *Philosophy and Science in the Scottish Enlightenment*, edited by P. Jones, 67–86. Edinburgh: John Donald, 1988.

Feyerabend, Paul. *Against Method.* London: NLB, 1975.

Feynman, Richard. *The Character of Physical Laws.* New York: Modern Library, 1994.

"Fictions." In *Encyclopaedia Britannica*, vol. 9, 218. Chicago: William Benton, 1962.

Fordyce, George. "An Account of an Experiment on Heat." *PT* 77 (1787): 310–17.

———. "An Account of Some Experiments on the Loss of Weight in Bodies on Being Melted or Heated." *PT* 75 (1785): 361–65.

———. "The Croonian Lecture on Muscular Motion." *PT* 78 (1788): 23–36.

Fox, Robert. *The Caloric Theory of Gases from Lavoisier to Regnault.* Oxford: Clarendon, 1971.

Franklin, Benjamin. *Benjamin Franklin's Experiments. A New Edition of Franklin's Experiments and Observations on Electricity*. Edited by I. B. Cohen. Cambridge, MA: Harvard University Press, 1941.

Freke, John. *An Essay to Shew the Cause of Electricity; and Why Some Things Are Non-electricable. . . . In a Letter to Mr. William Watson, F. R. S.* London, 1746.

Galilei, Galileo. *Dialogue Concerning the Two Chief World Systems.* Translated by S. Drake. Los Angeles: University of California Press, 1967.

———. *Dialogues Concerning Two New Sciences.* Translated by H. Crew and A. de Salvio. New York: McGraw-Hill, 1963.

Gell-Mann, Murray. *The Quark and the Jaguar: Adventures in the Simple and the Complex.* New York: W. H. Freeman, 1994.

Giere, Ronald N. *Explaining Science: A Cognitive Approach.* Chicago: University of Chicago Press, 1998.

Goldsmith, Oliver. *A Survey of Experimental Philosophy, Considered in Its Present State of Improvement.* 2 vols. London, 1766.

Goodricke, John. "A Series of Observations on, and a Discovery of, the Period of the Variation of the Light of the Bright Star in the Head of Medusa, Called Algol." *PT* 73 (1783): 474–82.

Gower, Barry. *Scientific Method: An Historical and Philosophical Introduction.* New York: Routledge, 1997.

Grattan-Guinness, I. "French *Calcul* and English Fluxions around 1800: Some Comparisons and Contrasts." *Jahrbuch Überblicke Mathematik*, 167–78. Mannheim: Bibliographisches Institut, 1986.

Grave, S. A. *The Scottish Philosophy of Common Sense.* Oxford: Clarendon Press, 1960.

Greenberg, John L. "Mathematical Physics in Eighteenth-Century France." *Isis* 77 (1986): 59–78.

Greene, John C. *Science, Ideology, and World View: Essays in the History of Evolutionary Ideas*. Berkeley: University of California Press, 1981.

Gregory, David. *The Elements of Astronomy, Physical and Geometrical.* 2 vols. London, 1715.

Guerlac, Henry. "Black, Joseph." In *Dictionary of Scientific Biography,* edited by C. C. Gillispie, vol. 2, 173–83. New York: Charles Scribner's Sons, 1970.

———. "Chemistry as a Branch of Physics: Laplace's Collaboration with Lavoisier." *Historical Studies in the Physical Sciences* 7 (1976): 193–276.

———. "Newton and the Method of Analysis." *Essays and Papers in the History of Modern Science*, 193–215. Balitmore: The Johns Hopkins University Press, 1977.

Hales, Stephen. *Vegetable Staticks. . . .* London, 1727.

Hall, A. Rupert Hall. *From Galileo to Newton 1630–1720*. London: Collins, 1963.

Hall, James. "Account of a Series of Experiments, Shewing the Effects of Compression in Modifying the Action of Heat." *TRSE* 6 (1805): 71–184.

———. "Experiments on Whinstone and Lava." *TRSE* 5:1 (1798): 43–75.

Hamilton, Hugh. *Four Introductory Lectures in Natural Philosophy*. London, 1774.

———. *Philosophical Essays on the Following Subjects: I. On the Principles of Mechanics. II. On the Ascent of Vapours. . . . III. Observations and Conjectures on the Nature of the Aurora Borealis, and the Tails of Comets*. Dublin, 1766.

Hankins, Thomas L. "Eighteenth-Century Attempts to Resolve the *Vis viva* Controversy." *Isis* 56 (1965): 281–97.

———. "Newton's 'Mathematical Way' a Century after the *Principia*." In *Some Truer Method: Reflections on the Heritage of Newton*, edited by F. Durham and R. D. Purrington, 89–112. New York: Columbia University Press, 1990.

———. "The Reception of Newton's Second Law of Motion in the Eighteenth Century." *Archives Internationales d'Histoire des Sciences* 20 (1967): 43–65.

———. *Science and the Enlightenment*. Cambridge: Cambridge University Press, 1985.

Harman, Peter. "Concepts of Inertia: Newton to Kant." In *Religion, Science, and Worldview: Essays in Honor of Richard S. Westfall*, edited by M. J. Osler and P. L. Farber, 109–33. Cambridge: Cambridge University Press, 1985.

———. *The Natural Philosophy of James Clerk Maxwell*. Cambridge: Cambridge University Press, 1998.

Harper, William L. "Reasoning from Phenomena: Newton's Argument for Universal Gravitation and the Practice of Science." *Action and Reaction: Proceedings of a Symposium to Commemorate the Tercentenary of Newton's Principia*, edited by P. Theeman and A. F. Seeff, 144–82. Newark: University of Delaware Press, 1993.

Harris, John. *Lexicon Technicum*. 5th edition. London, 1736.

Heilbron, J. L. *Electricity in the 17th and 18th Centuries: A Study of Early Modern Physics*. Los Angeles: University of California Press, 1979.

———. *Weighing Imponderables and Other Quantitative Science around 1800*. Supplement to *Historical Studies in the Physical and Biological Sciences.* Vol. 24, pt. 1. Berkeley: University of California Press, 1993.

Heimann, P. M. "Ether and Imponderables." In *Conceptions of Ether: Studies in the History of Ether Theories, 1740–1900*, edited by G. N. Cantor and M. J. S. Hodge, 61–83. Cambridge: Cambridge University Press, 1981.

———. " 'Geometry and Nature': Leibniz and Johann Bernoulli's Theory of Motion." *Centaurus* 21 (1977): 1–26.

———. " 'Nature is a Perpetual Worker': Newton's Aether and Eighteenth-Century Natural Philosophy." *Ambix* 20 (1973): 1–25.

———. "Newtonian Natural Philosophy and the Scientific Revolution." *History of Science* 11 (1973): 1–7.

Heimann, P. M., and J. E. McGuire. "Cavendish and the *Vis Viva* Controversy: A Leibnizian Postscript." *Isis* 62 (1970): 225–27.

———. "Newtonian Forces and Lockean Powers: Concepts of Matter in Eighteenth-Century Thought." *Historical Studies in the Physical Sciences* 3 (1971): 233–306.

Helsham, Richard. *A Course of Lectures in Natural Philosophy*. London, 1739.

Henley, William. "An Account of Some New Experiments in Electricity. . . ." *PT* 64 (1774): 389–431.

———. "Experiments and Observations in Electricity." *PT* 67 (1777): 85–145.

Herschel, William. "Astronomical Observations Relating to the Mountains of the Moon." *PT* 70 (1780): 507–20.

———. "Catalogue of the Second Thousand of New Nebulae and Clusters of Stars; with a Few Introductory Remarks on the Construction of the Heavens." *PT* 79 (1789): 212–52.

———. "On the Construction of the Heavens."*PT* 75 (1785): 213–66.

———. "On the Parallax of the Fixed Stars." *PT* 72 (1782): 82–111.

———. "On the Proper Motion of the Sun and Solar System; with an Account of Several Changes That Have Happened Among the Fixed Stars since the Time of Mr. Flamstead." *PT* 73 (1783): 247–83.

Hevly, Bruce. "Afterword: Reflections on Big Science and Big History." In *Big Science: The Growth of Large-Scale Research*, edited by P. Galison and B. Hevly, 355–63. Stanford, CA: Stanford University Press, 1992.

Hiebert, Erwin N. Comment. In *Critical Problems in the History of Science*, edited by M. Clagett, 394–96. Madison: University of Wisconsin Press, 1959.

———. *Historical Roots of the Principle of Energy Conservation*. Madison: State Historical Society of Wisconsin, 1962.

Higgins, Bryan. *Experiments and Observations Relating to Acetous Acid, Fixable Air, Dense Inflammable Air, Oils and Fuel; the Matter of Fire and Light. . . .* London, 1786.

Hillary, William. *The Nature, Properties, and Laws of Motion of Fire. Discovered and Demonstrated by Observations and Experiments.* London, 1760.

Home, R. W. "The Third Law in Newton's Mechanics." *British Journal for the History of Science* 4 (1968): 39–51.

Hooft, Gerard 't. "Questioning the Answers or Stumbling upon Good and Bad Theories of Everything." In *Physics and Our View of the World*, edited by J. Hilgevoord, 16–37. Cambridge: Cambridge University Press, 1994.

Horne, George. *A Fair, Candid, and Impartial Account of the Case between Sir Isaac Newton and Mr. Hutchinson.* . . . Oxford, 1753.

Horsley, Samuel. "Difficulties in the Newtonian Theory of Light, Considered and Removed." *PT* 60 (1770): 417–40.

———. "M. De Luc's Rules, for the Measurement of Heights by the Barometer, Compared with Theory. . . ." *PT* 64 (1774): 214–303.

Hume, David. *An Enquiry Concerning Human Understanding.* Edited by T. L. Beauchamp. Oxford: Oxford University Press, 1999.

———. *A Treatise of Human Nature.* Edited by L. A. Selby-Bigge. Oxford: Oxford University Press, 1955.

Humphreys, A. R. "The Literary Scene." In *From Dryden to Johnson*, vol. 4 of the Pelican Guide to English Literature, edited by B. Ford, 51–93. Baltimore: Penguin Books, 1963.

Hunter, John. "Of the Heat, etc, of Animals and Vegetables." *PT* 68 (1778): 7–49.

Hutchins, Thomas. "Experiments for Ascertaining the Point of Mercurial Congelation." *PT* 73 (1783): 303*–70*.

Hutton, Charles. "The Force of Fired Gun-Powder, and the Initial Velocities of Cannon Balls, Determined by Experiments; from Which Is Also Deduced the Relation of the Initial Velocity of the Weight of the Shot and the Quantity of Powder." *PT* 68 (1778): 50–85.

———. *A Mathematical and Philosophical Dictionary.* . . . 2 vols. London, 1795–96. New edition. 2 vols. London, 1815.

———. *Tracts on Mathematical and Philosophical Subjects.* . . . 3 vols. London, 1812.

Hutton, James. *A Dissertation upon the Philosophy of Light, Heat, and Fire. In Seven Parts.* Edinburgh, 1794.

———. *An Investigation of the Principles of Knowledge, and of the Progress of Reason, from Sense to Science and Philosophy.* 3 vols. Edinburgh, 1794.

Ingen-Housz, Jan. "Some Farther Considerations on the Influence of the Vegetable Kingdom on the Animal Creation." *PT* 72 (1782): 426–39.

Irvine, William. *Essays, Chiefly on Chemical Subjects.* Edited by W. Irvine, Jr. London, 1805.

Jones, William. *An Essay on the First Principles of Natural Philosophy: Wherein the Use of Natural Means, or Second Causes, in the Oeconomy of the Material World, Is Demonstrated from Reason, Experiments of Various Kinds, and the Testimony of Antiquity. . . . In Four Books.* . . . Oxford, 1762.

Jungnickel, Christa, and Russell McCormmach. *Cavendish, the Experimental Life.* Lewisburg, PA: Bucknell University Press, 1999.

———. *Intellectual Mastery of Nature: Theoretical Physics from Ohm to Einstein.* Vol. 1: *The Torch of Mathematics, 1800–1870.* Chicago: Chicago University Press, 1986.

Jurin, James. "An Inquiry into the Measure of the Force of Bodies in Motion: With a Proposal of an Experimentum Crucis, to Decide the Controversy about It." *PT* 43 (1745): 423–40.

Keir, James. "Experiments and Observations on the Dissolution of Metals and Acids, and Their Precipitations; with an Account of a New Compound Acid Menstruum, Useful in Some Technical Operations of Parting Metals." *PT* 80 (1790): 359–84.

Kepler, Johannes. *Epitome of Copernican Astronomy and Harmonies of the World.* Translated by C. G. Wallis. Amherst: Prometheus Books, 1995.

King, Edward. "An Attempt to Account for the Universal Deluge." *PT* 57 (1767): 44–57.

Kirwan, Richard. *An Essay on Phlogiston, and the Constitution of Acids.* 2d edition. London, 1789.

———. *An Estimate of the Temperature of Different Latitudes.* London, 1787.

———. "Thoughts on Magnetism." *TRIA* 6 (1796): 177–91.

Knight, Gowin. *An Attempt to Demonstrate, That All the Phaenomena in Nature May Be Explained by Two Simple Active Principles, Attraction and Repulsion: Wherein the Attractions of Cohesion, Gravity, and Magnetism, Are Shewn to Be One and the Same; and the Phaenomena of the Latter Are More Particularly Explained.* London, 1748.

———. "A Collection of the Magnetical Experiments Communicated to the Royal Society . . . in the Years 1746 and 1747." *PT* 44 (1747): 656–82.

Kuhn, Thomas S. "Mathematical versus Experimental Traditions in the Development of Physical Science." In *The Essential Tension: Selected Studies in Scientific Tradition and Change*, 31–65. Chicago: University of Chicago Press, 1977.

Laudan, L. L. "Thomas Reid and the Newtonian Turn of British Methodological Thought." In *The Methodological Heritage of Newton*, edited by R. E. Butts and J. W. Davis, 103–31. Toronto: University of Toronto Press, 1970.

Lavoisier, A. L., and P. S. Laplace. *Memoir on Heat. Read to the Royal Academy of Sciences, 28 June 1783 by Messrs. Lavoisier and De La Place.* Translated by H. Guerlac. New York: Neale Watson Academic Publications, 1982.

Leslie, P. Dugud. *A Philosophical Inquiry into the Cause of Animal Heat: With Incidental Observations on Several Physiological and Chymical Questions, Connected with the Subject.* London, 1778.

Leslie, John. *An Experimental Inquiry into the Nature, and Propagation of Heat.* London, 1804.

Lewis, William. *Commercium Philosophico-Technicum; or, The Philosophical Commerce of Arts: Designed as an Attempt to Improve Arts, Trades, and Manufactures.* London, 1763.

Locke, John. *An Essay Concerning Human Understanding.* Edited by A. C. Fraser. Vol. 2. Oxford: Clarendon Press, 1894.

———. *An Essay Concerning Human Understanding.* Abridged and edited by A. S. Pringle-Pattison. Oxford: Clarendon Press, 1950.

Lodwig, T. H., and W. A. Smeaton. "The Ice Calorimeter of Lavoisier and Laplace and Some of Its Critics." *Annals of Science* 31 (1974): 1–18.

Long, Roger. *Astronomy, in Five Books.* 2 vols. in 3. Cambridge, 1742–84.

Lorimer, J. "Description of a New Dipping-Needle." *PT* 65 (1775): 79–84.

Lovett, Richard. *Philosophical Essays, in Three Parts.* Worcester, 1766.

Ludlam, William. *Astronomical Observations Made in St. John's College, Cambridge, in the Years 1767 and 1768: With an Account of Several Astronomical Instruments.* Cambridge, 1769.

———. *Mathematical Essays.* 2d edition. Cambridge, 1787.

———. *The Rudiments of Mathematics; Designed for the Use of Students at the Universities: Containing an Introduction to Algebra, Remarks on the First Six Books of Euclid, the Elements of Plain Trigonometry.* Cambridge, 1785.

Lyon, John. *Experiments and Observations Made with a View to Point Out the Errors of the Present Received Theory of Electricity; and Which Tend in Their Progress to Establish a New System, on Principles More Conformable to the Simple Operations of Nature.* London, 1780.

Mach, Ernst. *Principien der Wärmelehre.* Leipzig, 1900.

Maclaurin, Colin. "An Account of a Book Intitled, *A Treatise of Fluxions, in Two Books,* by Colin McLaurin. . . ." *PT* 42 (1743): 325–63.

———. *An Account of Sir Isaac Newton's Philosophical Discoveries, in Four Books.* . . . London, 1748.

———. *A Treatise of Algebra, in Three Parts.* . . . London, 1748.

———. *A Treatise of Fluxions, in Two Books.* 2 vols. Edinburgh, 1742.

Madden, Edward H. "The Reidian Tradition: Growth of the Causal Concept." In *Nature and Scientific Method,* edited by D. O. Dahlstrom, 291–307. Washington, DC: The Catholic University of America Press, 1991.

Magellan, J. H. *Essai sur la Nouvelle Théorie du Feu Élémentaire, et de la Chaleur des Corps.* London, 1780.

Mairan, J. J. d'Ortous de. *Dissertation sur la Glace, ou Explication Physique de la Formation de la Glace, et de ses Divers Phénomènes.* Paris, 1749.

Maltese, Giulio. "Toward the Rise of the Modern Science of Motion: The Transition from Synthetical to Analytical Mechanics." In *History of Physics in Europe in the 19th and 20th Centuries.* Italian Physical Society Conference Proceedings, edited by F. Bevilacqua, vol. 42, 51–67. Bologna: Italian Physical Society, 1992.

Manuel, Frank. *A Portrait of Isaac Newton.* Cambridge, MA: Harvard University Press, 1968.

Martin, Benjamin. *A New and Comprehensive System of Mathematical Institutions, Agreeable to the Present State of the Newtonian Mathesis.* 2 vols. London, 1759–64.

———. *Philosophia Britannica; or, A New and Comprehensive System of the Newtonian Philosophy, Astronomy and Geography. In a Course of Twelve Lectures. . . .* 2 vols. Reading, 1747.

———. *The Philosophical Grammar; Being a View of the Present State of Experimental Physiology, or Natural Philosophy.* 5th edition. London, 1755.

Martine, George. *Essays and Observations on the Construction and Graduation of Thermometers, and on the Heating and Cooling of Bodies.* 2d edition. Edinburgh, 1772.

———. *Essays Medical and Philosophical.* 2d edition. Edinburgh, 1772.

McCormmach, Russell. "Henry Cavendish: A Study of Rational Empiricism in Eighteenth-Century Natural Philosophy." *Isis* 60 (1969): 293–306.

———. "Henry Cavendish on the Theory of Heat." *Isis* 79 (1988): 37–67.

McGuire, J. E. "Comment." *The* Annus Mirabilis *of Sir Isaac Newton.* Edited by R. Palter, 186–91. Cambridge, MA: London: MIT Press, 1970.

McKie, Douglas. "On Thos. Cochrane's MS. Notes of Black's Chemical Lectures, 1767–8." *Annals of Science* 1 (1936): 101–10.

McKie, Douglas, and Niels H. de V. Heathcote. *The Discovery of Specific and Latent Heats.* London: Arnold, 1935.

McMullin, Ernan. "Enlarging the Known World." In *Physics and Our View of the World*, edited by J. Hilgevoord, 79–113. Cambridge: Cambridge University Press, 1994.

———. *Newton on Matter and Activity.* Notre Dame: University of Notre Dame Press, 1978.

Melvill, Thomas. "Discourse Concerning the Cause of the Different Refrangibilities of the Rays of Light." *PT* 48 (1753): 261–70.

Merton, Robert K. *On the Shoulders of Giants: A Shandean Postscript.* San Diego: Harcourt Brace Jovanovich, 1985.

Michell, John. "Conjectures Concerning the Cause, and Observations upon the Phaenomena, of Earthquakes; Particularly of That Great Earthquake of the First of November, 1755, Which Proved So Fatal to the City of Lisbon, and Whose Effects Were Felt as Far as Africa, and More or Less throughout Almost All Europe." *PT* 51 (1760): 566–634.

———. "On the Means of Discovering the Distance, Magnitude, &c. of the Fixed Stars, in Consequence of the Diminution of the Velocity of Their Light, in Case Such a Diminution Should Be Found to Take Place in Any of Them, and Such Other Data Should Be Procured from Observations, as Would Be Farther Necessary for That Purpose." *PT* 74 (1784): 35–57.

———. *A Treatise of Artificial Magnets; in Which Is Shewn an Easy and Expeditious Method of Making Them, Superior to the Best Natural Ones; and Also, a Way of Improving the Natural Ones, and of Changing or Converting Their Poles. Directions Are Likewise Given for Making the Mariner's Needles in the Best Form, and for Touching Them Most Advantageously.* Cambridge, 1750. 2d edition. Cambridge, 1751.

Miller, David Philip. "The Revival of the Physical Sciences in Britain, 1815–40." *Osiris*, 2d ser., 2 (1986): 107–34.

———. "The Usefulness of Natural Philosophy: The Royal Society and the Culture of Practical Utility in the Later Eighteenth Century." *British Journal for the History of Science* 32 (1999): 185–201.

Morgan, Rev. Mr. [G. C.]. "Observations and Experiments on the Light of Bodies in a State of Combustion." *PT* 75 (1785): 190–212.

Morgan, William. "Electrical Experiments Made in Order to Ascertain the Non-conducting Power of a Perfect Vacuum, etc." *PT* 75 (1785): 272–78.

Morton, Alan Q. "Concepts of Power: Natural Philosophy and the Uses of Machines in Mid-Eighteenth-Century London." *British Journal for the History of Science* 28 (1995): 63–78.

Morton, Charles. "Observations and Experiments upon Animal Bodies, Digested in a Philosophical Analysis, or Inquiry into the Cause of Voluntary Muscular Motion." *PT* 47 (1751): 305–14.

Murdoch, Patrick. "Rules and Examples for Limiting the Cases in Which the Rays of Refracted Light May Be Reunited into a Colourless Pencil." *PT* 53 (1763): 173–94.

Nagel, Ernest. "Theory and Observation." In *Observation and Theory in Science*, edited by Ernest Nagel, Sylvain Bromberger, and Adolf Grünbaum. Baltimore: The Johns University Press, 1971.

Newton, Isaac. *Opticks; or A Treatise of the Reflections, Refractions, Inflections and Colours of Light.* 4th edition of 1730. Reprint. New York: Dover, 1952.

———. *Sir Isaac Newton's Mathematical Principles of Natural Philosophy and His System of the World.* Translated by A. Motte in 1729. Edited by F. Cajori. 2 vols. Los Angeles: University of California Press, 1962.

Nicholson, William. *A Dictionary of Chemistry. . . .* 2 vols. in 1. London, 1795.

———. *The First Principles of Chemistry.* London, 1790.

———. *An Introduction to Natural Philosophy.* 2 vols. London, 1782.

Oliver, J. "William Borlase's Contributions to Eighteenth-Century Meteorology and Climatology." *Annals of Science* 25 (1969): 275–317.

Olson, Richard. *Scottish Philosophy and British Physics 1750–1880: A Study in the Foundations of the Victorian Scientific Style.* Princeton, NJ: Princeton University Press, 1975.

The Oxford Universal Dictionary on Historical Principles. Revised and edited by C. T. Onions. 3d edition. Oxford: Clarendon Press, 1955.

Parkinson, Thomas. *A System of Mechanics, Being the Substance of Lectures upon That Branch of Natural Philosophy*. Cambridge, 1785.

Passmore, John. "Hume, David." In *Dictionary of Scientific Biography*, edited by C. C. Gillispie, vol. 6, 555–60. New York: Charles Scribner's Sons, 1972.

Penrose, Roger, et al. *The Large, the Small and the Human Mind*. Cambridge: Cambridge University Press, 1997.

"Physics." In *Encyclopaedia Britannica*, vol. 3, 478. Edinburgh, 1777.

"Physiology." *Encyclopaedia Britannica*, vol. 3, 478. Edinburgh, 1777.

Pigott, Edward. "Observations of a New Variable Star." *PT* 75 (1785): 127–36.

Playfair, John. *Illustrations of the Huttonian Theory of the Earth*. Edinburgh, 1802.

———. *Outlines of Natural Philosophy, Being Heads of Lectures Delivered in the University of Edinburgh*. 2 vols. in 1. Edinburgh, 1812–14.

———. *The Works of John Playfair*. Edited by J. G. Playfair. 4 vols. Edinburgh, 1822.

Porter, Roy. *Enlightenment: Britain and the Creation of the Modern World*. London: Allen Lane, 2000.

"Preface." *TRIA* 1 (1787): ix–xvii.

Priestley, Joseph. "An Account of Further Discoveries in Air." *PT* 65 (1775): 384–94.

———. *Disquisitions Relating to Matter and Spirit. To Which Is Added, the History of the Philosophical Doctrine Concerning the Origin of the Soul, and the Nature of Matter*. . . . London, 1777.

———. "Experiments Relating to Phlogiston, and the Seeming Conversion of Water into Air." *PT* 73 (1783): 398–434.

———. *The History and Present State of Discoveries Relating to Vision, Light, and Colours*. 2 vols. London, 1772.

———. *The History and Present State of Electricity, with Original Experiments*. London, 1767.

———. *A Scientific Autobiography of Joseph Priestley (1733–1804): Selected Scientific Correspondence*. Edited by R. E. Schofield. Cambridge, MA: MIT Press, 1966.

Pringle, John. "Some Remarks upon the Several Accounts of the Fiery Meteor (Which Appeared on Sunday the 26th of November, 1758), and upon Other Such Bodies." *PT* 51 (1760): 259–74.

Ravetz, J. "The Representation of Physical Quantities in Eighteenth-Century Mathematical Physics." *Isis* 52 (1961): 7–20.

Reid, Thomas. "An Essay on Quantity; Occasioned by Reading a Treatise in Which Simple and Compound Ratios Are Applied to Virtue and Merit, by the Rev. Mr. Reid; Communicated in a Letter from the Rev. Henry Miles D.D. and F.R.S. to Martin Folkes Esq; Pr.R.S." *PT* 45 (1748): 505–20.

———. *The Works of Thomas Reid, D.D.* Edited by W. Hamilton. 7th edition. Vol. 2. Edinburgh, 1872.

Rhodes, Richard. *The Making of the Atomic Bomb.* New York: Toronto: Simon & Schuster, 1986.

Robins, Benjamin. "An Account of a Book Intitled, New Principles of Gunnery, Containing the Determination of the Force of Gunpowder; and an Investigation of the Resisting Power of the Air to Swift and Slow Motions . . . as Far as the Same Relates to the Force of Gunpowder." *PT* 42 (1743): 437–56.

———. *A Discourse Concerning the Nature and Certainty of Sir Isaac Newton's Methods of Fluxions and of Prime and Ultimate Ratios.* In *Mathematical Tracts* 1: 7–77.

———. *Mathematical Tracts of the Late Benjamin Robins, Esq.* Edited by J. Wilson. Vol. 1. London, 1761.

———. *New Principles of Gunnery.* London, 1742.

Robinson, Bryan. *A Dissertation on the Aether of Sir Isaac Newton.* Dublin, 1743.

———. *Sir Isaac Newton's Account of the Aether, with Some Additions by Way of Appendix.* Dublin, 1745.

Robison, John. "On the Motion of Light, as Affected by Refracting and Reflecting Substances, Which Are Also in Motion." *TRSE* 2:2 (1788): 83–111.

———. "Physics." In *Encyclopaedia Britannica*, vol. 14, 637–59. 3d edition. Edinburgh, 1797.

———. *A System of Mechanical Philosophy. . . .* Edited by D. Brewster. 4 vols. Edinburgh, 1822.

Roche, John. "Newton's *Principia*," In *Let Newton Be!*, edited by J. Fauvel, R. Flood, M. Shortland, and R. Wilson, 43–62. Oxford: Oxford University Press, 1988.

Roebuck, John. "Experiments on Ignited Bodies." *PT* 66 (1776): 509–12.

Rossi, Arcangelo. "Kantism, Phenomenalism, Reductionism and the Emergence of Theoretical Physics in the 19th Century." In *History of Physics in Europe in the 19th and 20th Centuries*, edited by F. Bevilaqua, 279–85. Conference Proceedings of the Italian Physical Society. Vol. 42. Bologna: Italian Physical Society, 1992.

Rouse Ball, W. W. *A History of the Study of Mathematics at Cambridge.* Cambridge: Cambridge University Press, 1889.

Rowning, John. *A Compendious System of Natural Philosophy: With Notes Containing Mathematical Demonstrations, and Some Occasional Remarks.* 3d edition. London, 1738.

Rutherforth, Thomas. *A System of Natural Philosophy; Being a Course of Lectures in Mechanics, Optics, Hydrostatics, and Astronomy, Which Are Read in St. Johns College Cambridge. . . .* 2 vols. Cambridge, 1748.

Salmon, Wesley C. *Scientific Explanation and the Causal Structure of the World.* Princeton, NJ: Princeton University Press, 1984.

Saussure, H. B. de. *Voyages dans les Alpes. . . .* Vol. 2. Neuchâtel, 1786.

Schaffer, Simon. "Natural Philosophy." In *The Ferment of Knowledge*, edited by G. S. Rousseau and R. S. Porter, 55–91. Cambridge: Cambridge University Press, 1980.

———. "Natural Philosophy and Public Spectacle in the Eighteenth Century. *History of Science* 21 (1983): 1–43.

———. "Scientific Discoveries and the End of Natural Philosophy." *Social Studies of Science* 16 (1986): 387–420.

Scheele, Carl Wilhelm. *Chemical Observations and Experiments on Air and Fire.* Translated by J. R. Forster, with notes by Richard Kirwan. London, 1780.

Schofield, Robert E. *Mechanism and Materialism: British Natural Philosophy in an Age of Reason.* Princeton, NJ: Princeton University Press, 1970.

"Science." In Encyclopaedia Britannica, vol. 16, 705. 3d edition. Edinburgh, 1797.

Scott, E. L. "Richard Kirwan, J. H. de Magellan, and the Early History of Specific Heat." *Annals of Science* 38 (1981): 141–53.

Scott, J. F. "Maclaurin, Colin." In *Dictionary of Scientific Biography*, edited by C. C. Gillispie, vol. 8, 609–12. New York: Charles Scribner's Sons, 1973.

Shapere, Dudley. "The Philosophical Significance of Newton's Science." In *The Annus Mirabilis of Sir Isaac Newton*, edited by R. Palter, 285–99. Cambridge, MA: MIT Press, 1970.

Shepherd, Anthony. *The Heads of a Course of Lectures in Experimental Philosophy Read at Christ College.* Cambridge, [1770].

Sibum, Heinz Otto. "Reworking the Mechanical Value of Heat: Instruments of Precision and Gestures of Accuracy in Early Victorian England." *Studies in the History and Philosophy of Science* 26 (1995): 73–106.

Simpson, Thomas. *The Doctrine and Application of Fluxions.* 2 vols. London, 1750.

———. *Miscellaneous Tracts. . . .* London, 1757.

Smeaton, John. "An Experimental Inquiry Concerning the Natural Powers of Water and Wind to Turn Mills, and Other Machines, Depending on Circular Motion." *PT* 51 (1760): 100–74.

Smith, Adam. "The Principles Which Lead and Direct Philosophical Inquiries; Illustrated by the History of Astronomy." In *The Whole Works of Adam Smith, LL.D. F.R.S. &c.*, vol. 5, 1–80. New edition. London, 1822.

Smith, Crosbie. " 'Mechanical Philosophy' and the Emergence of Physics in Britain: 1800–1850." *Annals of Science* 33 (1976): 3–29.

———. See Maurice Crosland.

Smith, Robert. *A Compleat System of Optics in Four Books, viz. a Popular, a Mathematical, a Mechanical, and a Philosophical Treatise. To Which Are Added Remarks upon the Whole.* 2 vols. Cambridge, 1738.

———. *Harmonics, or The Philosophy of Musical Sounds.* Cambridge, 1749. Reprint, New York: Da Capo, 1966.

Sorrenson, Richard. "Towards a History of the Royal Society in the Eighteenth Century." *Notes and Records of the Royal Society of London* 50 (1996): 29–46.

Steele, Brett D. "Muskets and Pendulums: Benjamin Robins, Leonhard Euler, and the Ballistics Revolution." *Technology and Culture* 35 (1994): 348–82.

Steffens, Henry John. *The Development of Newtonian Optics in England.* New York: Science History Publications, 1977.

Stewart, Dugald. *The Collected Works of Dugald Stewart, Esq., F.R.SS.* 11 vols. Edited by W. Hamilton. Edinburgh, 1854–1860.

Stewart, Larry. "Seeing Through the Scholium: Religion and Reading Newton in the Eighteenth Century." *History of Science* 34 (1996): 123–64.

Straub, Hans. "Bernouilli, Daniel." In *Dictionary of Scientific Biography*, edited by C. C. Gillispie, vol. 2, 36–46. New York: Charles Scribner's Sons, 1970.

Symmer, Robert. "New Experiments and Observations Concerning Electricity." *PT* 51 (1760): 340–93.

Taylor, Brook. "An Account of an Experiment, Made to Ascertain the Proportion of the Expansion of the Liquor in the Thermometer, with Regard to Degrees of Heat." *PT* 32 (1723): 291.

Thompson, Benjamin. *The Complete Works of Count Rumford.* 4 vols. Boston, 1870–75.

———. "Experiments to Determine the Force of Fired Gunpowder." *PT* 87 (1797): 222–92.

———. "An Inquiry Concerning the Source of the Heat Which Is Excited by Friction." *PT* 88 (1798): 80–102.

———. "An Inquiry Concerning the Weight Ascribed to Heat." *PT* 89 (1799): 179–94.

———. "New Experiments upon Gun-Powder, with Occasional Observations and Practical Inferences; to Which Are Added, an Account of a New Method of Determining the Velocities of All Kinds of Military Projectiles, and the Description of the Very Accurate Eprouvette for Gun-Powder." *PT* 71 (1781): 229–328.

Thrower, N. J., ed. *Standing on the Shoulders of Giants: A Longer View of Newton and Halley. Essays Commemorating the Tercentenary of Newton's* Principia *and the 1985–1986 Return of Comet Halley*. Los Angeles: University of California Press, 1990.

Truesdell, C. "A Program toward Rediscovering the Rational Mechanics of the Age of Reason." *Archive for History of Exact Sciences* 1 (1960): 1–36.

———. *Rational Fluid Mechanics, 1687–1765*. Editor's Introduction, *Leonhardi Euleri Opera omnia*, 2d ser. Vol. 12, pt. 1, ix–cxxv. Zurich: Orell Füssli, 1954.

———. *The Rational Mechanics of Flexible or Elastic Bodies 1638–1788*. Editor's Introduction, *Leonhardi Euleri Opera omnia.* 2d ser. Vol. 11, pt. 2. Zurich: Orell Füssli, 1960.

———. "Reactions of Late Baroque Mechanics to Success, Conjecture, Error, and

Failure in Newton's *Principia*." In *The* Annus Mirabilis *of Sir Isaac Newton 1666–1966*, edited by R. Palter, 192–232. Cambridge, MA: MIT Press, 1970.

Ussher, Henry. "Account of the Observatory Belonging to Trinity College, Dublin." *TRIA* 1 (1787): 3–22.

Vagliente, Victor N. See Auguste Claude Boissonnard.

Vince, Samuel. *The Heads of a Course of Lectures on Experimental Philosophy; Comprising All the Fundamental Principles in Mechanics, Hydrostatics, and Optics; with an Explanation of the Construction and Use of All the Principal Instruments in Astronomy; Together with Magnetism and Electricity*. Cambridge, 1795.

———. "On the Motion of Bodies Affected by Friction." *PT* 75 (1785): 165–89.

———. *A Plan of a Course of Lectures on the Principles of Natural Philosophy.* Cambridge, 1793.

Walmesley, Charles. "Two Essays Addressed to the Rev. James Bradley, D.D." *PT* 49 (1755): 700–59.

Watson, Richard. "An Account of an Experiment Made with a Thermometer, Whose Bulb Was Painted Black, and Exposed to the Direct Rays of the Sun." *PT* 63 (1773): 40–41.

———. "Observations on the Sulphur Wells at Harrowgate, Made in July and August, 1785." *PT* 76 (1786): 171–88.

Watson, William. "A Collection of the Electrical Experiments Communicated to the Royal Society." *PT* 45 (1748): 49–120.

———. "A Sequel to the Experiments and Observations Tending to Illustrate the Nature and Properties of Electricity." *PT* 44 (1747): 704–49.

Wedgwood, Josiah. "Additional Observations on Making a Thermometer for Measuring the Higher Degrees of Heat." *PT* 76 (1786): 390–408.

———. "An Attempt to Compare and Connect the Thermometer for Strong Fire, Described in Vol. LXXII of the Philosophical Transactions with the Common Mercurial One." *PT* 74 (1784): 358–84.

Weinberg, Steven. *Dreams of a Final Theory.* New York: Pantheon Books, 1992.

Westfall, Richard S. *Science and Religion in Seventeenth-Century England.* Ann Arbor: University of Michigan Press, 1973.

White, W. "Experiments upon Air; and the Effects of Different Kinds of Effluvia upon It; Made at York." *PT* 68 (1778): 194–220.

Wilber, K., ed. *Quantum Questions: Mystical Writings by the World's Great Physicists.* Boston: Shambhala, 1985.

Wilde, C. B. "Hutchinsonianism, Natural Philosophy and Religious Controversy in Eighteenth Century Britain." *History of Science* 18 (1980): 1–24.

Wilkinson, Charles Henry. *An Analysis of a Course of Lectures on the Principles of Natural Philosophy, to Which Is Prefixed, An Essay on Electricity, with a View of Explaining the Phenomena of the Leyden Phial, etc. on Mechanical Principles.* London, 1799.

Wilson, Alexander. "An Answer to the Objections Stated by M. De la Lande, in the Memoirs of the French Academy in the Year 1776, against the Solar Spots Being Excavations in the Luminous Matter of the Sun, Together with a Short Examination of the Views Entertained by Him upon That Subject." *PT* 73 (1783): 144–68.

———. "Observations on the Solar Spots." *PT* 64 (1774): 1–30.

———. *Thoughts on General Gravitation, and Views Thence Arising as to the State of the Universe.* London, 1777.

Wilson, Benjamin. "Experiments on the Tourmalin." *PT* 51 (1759): 308–39.

———. *A Short View of Electricity.* London, 1780.

Wilson, George. *The Life of the Honourable Henry Cavendish. . . .* London, 1851.

Wilson, Patrick. "An Experiment Proposed for Determining, by the Aberration of the Fixed Stars, Whether the Rays of Light, in Pervading Different Media, Change Their Velocity According to the Law Which Results from Sir Isaac Newton's Ideas Concerning the Cause of Refraction; and for Ascertaining Their Velocity in Every Medium Whose Refractive Density Is Known." *PT* 72 (1782): 58–71.

Wolf, A. *A History of Science, Technology and Philosophy in the 16th and 17th Centuries.* 2d edition. Edited by D. McKie. Vol. 1. New York: Harper and Brothers, 1959.

———. *A History of Science, Technology, and Philosophy in the 18th Century.* 2d edition. Edited by D. McKie. 2 vols. New York: Harper and Brothers, 1961.

Wolff, Stefan L. "Origins of Theoretical Physics in Germany in 19th Century." In *I Beni Culturali Scientifici nella Storia e Didattica. Atti del Convego del 14–15 dicembre 1990*, 162–76. Pavia: Università degli Studi di Pavia, 1990.

Wollaston, W. H. "The Bakerian Lecture on the Force of Percussion." *PT* 96 (1806): 13–22.

Wordsworth, Christopher. *Scholae Academicae: Some Account of the Studies at the English Universities in the Eighteenth Century.* Cambridge, 1877.

Wright, Thomas. *An Original Theory or New Hypothesis of the Universe, Founded upon the Laws of Nature, and Solving by Mathematical Principles the General Phaenomena of the Visible Creation; and Particularly the Via Lactae. . . .* London, 1750.

Yost, Robinson M. "Pondering the Imponderable: John Robison and Magnetic Theory in Britain (c. 1775–1805)." *Annals of Science* 56 (1999): 143–74.

Young, Matthew. *An Analysis of the Principles of Natural Philosophy.* Dublin, 1800.

———. "Demonstration of Newton's Theorem for the Correction of Spherical Errors in the Object Glasses of Telescopes." *TRIA* 4 (1792): 171–75.

Young, Robert. *An Essay on the Powers and Mechanism of Nature; Intended by a Deeper Analysis of Physical Principles, to Extend, Improve, and More Firmly Establish, the Grand Superstructure of the Newtonian System.* London, 1788.

Young, Thomas. "The Bakerian Lecture. On the Theory of Light and Colours." *PT* 92 (1802): 12–48.

———. *A Course of Lectures on Natural Philosophy and the Mechanical Arts.* 2 vols. London, 1807.

———. "Outlines of Experiments and Inquiries Respecting Sound and Light." *PT* 90 (1800): 106–50.

———. *A Syllabus of a Course of Lectures on Natural and Experimental Philosophy.* London, 1802.

INDEX